Pharmacum-Pharmacum and Pharmacum-Viands Interrelationment on the Pilotage Pathosis

Klemens Wiese

Pharmacum-Pharmacum and Pharmacum-Viands Interrelationment on the Pilotage Pathosis

First Edition August 2025

Written by Klemens Wiese

CONTENTS

1. INTRODUCTION

Hyperlipidaemia[1] is a condition in which blood plasma contains high levels of lipids and/or lipoproteins. This can be primary due to genetic factors or secondary which may be due to underlying things such as diabetes, hypothyroidism, nephrotic syndrome, and alcohol, as well as the dietary intake.

Lipids such as cholesterol and triglycerides are insoluble in plasma. Circulating lipid is carried in lipoproteins that transport the lipid to various tissues for energy use, lipid deposition, steroid hormone production, and bile acid formation. Most people who have Hyperlipidaemia experience no symptoms. Abnormalities in lipoprotein metabolism are a major predisposing factor to atherosclerosis, increasing risk for Coronary heart disease (CHD). Cardiovascular disease due to atherosclerosis of the arterial vessel wall and to thrombosis is the foremost cause of pre-mature mortality and of disability-adjusted life years, and is also increasingly common in developing countries. The main clinical entities are coronary artery disease, ischaemic stroke, and peripheral arterial disease[2].

Coronary heart disease[3] is the most common cause of death worldwide. According to WHO, an estimated 7.2 million people died from CHD in 2008, representing approximately 12% of deaths worldwide, while in the year 2030 it is estimated that 23.6 million people will die from cardiovascular disease.

World health organization[4] has drawn attention to the fact that coronary heart disease as our modern EPIDEMIC i.e., a disease that affects population not an unavoidable attribute of aging. In India, the burden of ischemic heart disease is increasing every year, because of consequence of exposure to risk factors likes inappropriate nutrition, sedentary life, smoking, obesity etc. It is expected to be the single most important cause of death in India by year 2015.

The rate of coronary heart disease has risen from 4% to 11% in past five decades. Recently World Health Organization (WHO) has declared that by 2020, 60% of cardiovascular cases will be of Indian origin[5].

Hence control[6] of Hyperlipidaemia can prevent recurrent attack of this life threatening diseases. The drugs reduce blood cholesterol levels by 25-35% and cause a 35-45% reduction in risk of ischaemic heart disease.

Hyperlipidaemia characterized by increased levels of total cholesterol, LDL-C and triglycerides, is a major modifiable risk factor in primary and secondary prevention of coronary artery disease. However the term dyslipidaemia is preferred to hyperlipidaemia because low levels of plasma HDL-C levels can be harmful.

Cholesterol is a soft waxy steroid. In spite of its seemingly bad influence on the body, it is an essential constituent of cell membranes and is crucial for normal body functions, including formation of bile acids, steroid hormones, vitamin D, sex hormones such as androgens and estrogens and metabolism of fat soluble vitamins (A, D, E, K). It plays an important role in permeability of cell membranes and in the prevention of crystallization of hydrocarbons. Furthermore, cholesterol, along with other plasma lipids such as triglycerides and phospholipids, provide energy and assist in maintenance of body temperature.

The daily consumption of food provides part of the required cholesterol and 20% to 25% is manufactured by the liver. The remainder is synthesized by the intestine, adrenal glands, reproductive organs and other tissues. Cholesterol is found in abundance in egg yolk, various oils, fats, nerve tissue of the spinal cord, brain and kidneys. A normal blood value of cholesterol is healthy. However, as will be discussed, elevated blood levels are harmful and have been associated with cardiovascular diseases. Abnormally elevated lipids in the blood stream allow cholesterol, particularly low density lipoprotein cholesterol (LDL-C), to be deposited within the arterial wall of large and medium sized arteries as atherosclerotic plaques. These cause obstruction of the arteries, and depending on the extent of obstruction, may contribute to hypertension, reduction in the amount of oxygenated blood that reaches the heart and in increasing the risk of coronary heart disease

(CHD), myocardium infarction and cerebral arterial diseases. Despite improvement in lifestyle (diet, exercise and weight reduction) and the use of cholesterol lowering drugs, CHD and stroke remain as major causes of death.

The liver of an individual with average frame and weight synthesizes about 1000 mg of cholesterol daily. The total cholesterol content of the body is approximately 35 grams. The average dietary intake of an adult is from 200 mg to 300 mg per day. Normally, the body tends to compensate for cholesterol by reducing the manufactured quantity. Once synthesized by the liver, cholesterol is transferred via the bile into the intestinal tract. About 50% of excreted cholesterol is reabsorbed by the digestive system and pumped back into circulation. This cholesterol recycling is continuous in nature. Plant sterol, when included in the diet, tends to compete with cholesterol absorption, resulting in reducing cholesterol blood levels. Biosynthesis and regulation of cholesterol depends on cholesterol blood level. The higher the intake of cholesterol, the less endogenous production, and the opposite is true.

1.1 CHOLESTEROL

Lipids consist of a number of different chemicals: free fatty acids, triglycerides, sterols (cholesterol and cholesterol esters), and phospholipids (phosphoric acid esters of lipids). Triglycerides exist in nature as solids (fats) or liquids (oils). This depends on room temperature, the length of the fatty acid chain and the extent of their hydrogen ion saturation. Triglycerides with short fatty acids and/or unsaturated fatty acids exist as liquids at room temperature (plant oils such as olive oil). Triglycerides with long fatty acid chains and/or saturated fatty acids exist in the solid form at room temperature (animal fats such as butter).

Lipids in blood circulation do not exist in the free form but rather as complexes (lipoproteins). To facilitate their transport, lipids bind to plasma protein such as globulin or albumin to form these complexes.

Very Low Density Lipoprotein (VLDL)

VLDL cholesterol is produced by the liver and is made up of 50% to 65% glycerides and 20% to 30% cholesterol. It is responsible for transporting triglycerides synthesized in the liver to adipose and muscular tissue. What remains of VLDL is broken down to LDL.

Low Density Lipoprotein Cholesterol (LDL)

LDL, often referred to as the "bad cholesterol," consists of a predominantly cholesterol inner core. It forms as a result of the breakdown of the metabolites of VLDL. It is made up of 51% to 58% of cholesterol and 4% to 8% of triglycerides. It makes up about 60% to 75% of all plasma cholesterol. Its main function is to deliver cholesterol from the liver cells. If large quantities of LDL are carried and no new LDL receptors are formed, the LDL absorption will be diminished and a harmful buildup of LDL will take place which may increase the risk of CHD. A 25% reduction in plasma LDL level may reduce the occurrence of CHD by 50%.

High Density Lipoprotein (HDL)

This cholesterol, which is known as "good cholesterol," tends to prevent arterial disease from occurring as it takes cholesterol away from the cells and back to the liver. Once in the liver, it may breakdown or be excreted from the body as waste. It is the smallest and densest of lipoproteins. High density lipoprotein constituents include 18% - 25% cholesterol and 2% - 7% triglycerides. It contributes approximately 20% - 30% of total cholesterol in the blood stream. The main function of HDL is to transport cholesterol from the body tissue to the liver where it is broken down and excreted in the bile. Thus, accumulation of cholesterol is prevented. The amount of transported cholesterol is about 25% of the cholesterol in plasma. Unlike high LDL and VLDL blood levels, high HDL blood levels reduce the risk of incidence of CHD. It has been shown that healthy diet and physical exercise tend to elevate HDL blood level.

Other lipids that play a role in healthy arteries are chylomicrons and triglycerides.

Chylomicrons are lipoprotein particles that transport dietary lipids from the intestines to other parts of the body. It is manifested by the absorptive cells of the small intestine. Chylomicrons are composed of from 85% - 92% triglycerides, 6% - 12% phospholipids, 1% - 3% cholesterol, and 1% - 2% protein. These particles deal with the transport of dietary lipids to the liver, adipose, cardiac, and skeletomuscular tissues. The majority of chylomicrons are deactivated in the blood by the enzyme lipoprotein lipase within 12 to 14 hours.

Triglycerides are a combination of glycerol and three different fatty acids. Most fats in the blood exist as triglycerides in association with cholesterol. Since triglycerides are insoluble in water, they are transported in combination with protein. Blood triglycerides originate either from fat present in diet or may be manufactured from carbohydrates stored in the body. Excess caloric intake is converted into triglycerides and stored in fat cells until such a time when food intake is reduced or not enough to provide energy. In such circumstances and under hormonal influences, triglycerides are released from fat cells as energy sources.

1.2 HYPERLIPIDAEMIA

Dyslipidemia[7] refers to the derangements of one or many of the lipoproteins; elevations of total cholesterol, low density lipoprotein (LDL) cholesterol and/or triglycerides, or low levels of high-density lipoprotein (HDL) cholesterol while elevation of lipoproteins alone is labeled as 'Hyperlipidemia'. Hyperlipoproteinaemias are disorders of metabolism in which one or more of the plasma lipoproteins are increased. Hypothyroidism, which is not infrequent cause of secondary lipoprotein disorders, often manifests with elevated LDL cholesterol, triglycerides, or both.

Estrogens[8] can elevate plasma triglycerides and HDL cholesterol levels, probably because of increases in both hepatic VLDL and apo-AI production. In postmenopausal women, estrogens may reduce LDL cholesterol by up to 15 percent.

Male sex hormones and anabolic steroids can increase hepatic lipase activity and have been used in the treatment of hypertriglyceridemia in men. Growth hormone can reduce LDL cholesterol and augment HDL cholesterol but is not recommended in the treatment of lipoprotein disorders.

The most frequent secondary cause of dyslipoproteinemia is probably the constellation of metabolic abnormalities seen in subjects with the metabolic syndrome. The finding of increased visceral fat (abdominal obesity), elevated blood pressure, and impaired glucose tolerance often clusters with increased plasma triglycerides and a reduced HDL cholesterol level. These are the major components of the metabolic syndrome.

Familial lipodystrophy (complete or partial) may be associated with increased VLDL secretion. Dunnigan lipodystrophy, a genetic disorder with features of the metabolic syndrome, is caused by mutations within the Lamin A/C gene and is associated with limb-girdle fat atrophy. Excess plasma triglycerides often accompany glycogen storage disorders.

1.2.1 Causes of Hyperlipidaemia

Hyperlipidaemia is due to genetic and environmental factors[9], including:

- Presence of diseases that tend to increase LDL blood levels. Such diseases include, but are not limited to: diabetes, hypertension, hypertriglyceridemia, kidney and liver diseases.
- Family history of developing CHD or CVA early in their lives (under 55 for brother and father and under 65 years of age for mother and sister). Likewise, family history of Hyperlipidaemia early in life will increase the risk of developing Hyperlipidaemia.
- Gender: Men have a greater chance of developing Hyperlipidaemia than women.
- Age: As a person becomes older, so does the chance for developing atherosclerosis and Hyperlipidaemia.
- Many foods such as eggs, butter, liver, kidneys, and certain sea foods contain cholesterol in amounts that will not drastically change cholesterol blood levels. Other foods, especially if consumed in relatively large quantities and frequently, can detrimentally affect cholesterol and triglyceride blood levels. Red

meat, many cheeses, creamy cakes, ice cream, sausages and hot dogs have high contents of saturated fats and may affect the outcome of cholesterol blood concentration.

- Sedentary lifestyle: It has been shown that non-vigorous physical activity tends to reduce HDL and elevate LDL blood levels.
- Bodyweight: Individuals who maintain normal bodyweight which is adequate for their frame and age are less likely to have high LDL and lower HDL levels than overweight or obese individuals. In general, overweight individuals do not follow a healthy diet.
- Smoking: It has been reported that smoking contributes to about 40, 00,000 deaths annually. In addition to contributing to cancer and cardiovascular diseases, it detrimentally affects the levels of LDL and HDL. Cigarette smoking decreases HDL level while it elevates LDL.
- Alcoholic Beverages: Persons who regularly consume large quantities of alcoholic beverages exhibit high LDL and low HDL levels. Cholesterol blood level is normally unaffected in people who do not drink or who drink in moderation.

1.3 RISKS OF HYPERLIPIDAEMIA

1.3.1 Atherosclerosis

Atherosclerosis is a specific form of arteriosclerosis affecting primarily the intima of the large and medium sized muscular arteries and is characterized by fibro fatty plaques are atheroma. The term atherosclerosis is derived from 'athero' (meaning porridge) referring to the soft lipid rich material in the centre of atheroma and 'sclerosis' (scarring) referring to connective tissue in the plaques.

Atherosclerosis and its relevant vascular events including cardiovascular disease (CVD), stroke and peripheral arterial disease (PAD) have become a leading cause of disability and mortality in modern society.

Atherogenesis in humans typically occurs over a period of many years, usually many decades. After a generally prolonged "silent" period, atherosclerosis may become clinically manifest. Atherosclerosis involves the build up of plaque composed of variable amounts of LPS, extracellular matrix(collagen, proteoglycans, glycosaminoglycans), calcium, vascular smooth muscle cells, inflammatory cells (chiefly monocytes-derived macrophages, T-lymphocytes, mast cells, dendritic cells). Atherosclerosis represents chronic inflammatory response to vascular injury caused by variety of agents.

Age

Death rates from Ischaemic heart disease rise with each decade. Between ages 40 and 60 the incidence of myocardial infarction increases fivefold.

Sex Men are much more prone to atherosclerosis and its consequences than women. After menopause, the incidence of atherosclerosis related diseases increases, probably due to a decrease in natural estrogen levels.

Risk factors for atherosclerosis

Major Non Modifiable:	**Lesser, Uncertain or non-quantitated**
Increasing age	Obesity
Male gender	Physical inactivity
Family history	Stress (Type A personality)
Genetic abnormalities	High carbohydrate intake
	Postmenopausal
	Estrogen deficiency

Potentially (modifiable) controllable:

Hyperlipidaemia	Alcohol
Hypertension	Lipoprotein [LP(a)]
Cigarette smoking	Hardened unsaturated fat intake
Diabetes mellitus	Chlamydia pneumonia

Obesity

If the person is overweight by 20% or more, is associated with increased risk. Obesity induces hypertension, diabetes, hypertriglyceridaemia and decreased HDL-C.

Genetics

The well-established familial predisposition to atherosclerosis and IHD is most likely polygenic. Most commonly, the genetic propensity relates to familial clustering of other risk factors, such as hypertension or diabetes, while less commonly involves well defined hereditary genetic derangements in lipoprotein metabolism that result in excessively high blood lipid levels.

Diet

Individuals with elevated levels of blood cholesterol, have a high incidence of atherosclerosis. Dietary fats containing saturated, fatty acids are associated with high levels of total plasma cholesterol and LDL-C and result inincreased risk of coronary artery disease. Polyunsaturated fatty acids and monounsaturated fatty acids lower plasma cholesterol.

Alcohol

Regular intake of small to moderate amounts has been found to raise HDL-C levels and decrease LDL-C oxidation. Increased consumption of alcohol causes increased synthesis of VLDL-C, with consequent hypertriglyceridaemia.

Hypertension

Hypertension is a major risk factor for atherosclerosis at all ages. Anti-hypertensive therapy reduces the incidence of atherosclerosis related diseases, particularly stroke and ischaemic heart disease.

Cigarette smoking

Cigarette smoking is a well-established risk factor. Smoking one or more packs of cigarettes per day for several years increases the death rate from IHD up to 200%.

Diabetes mellitus

Diabetes mellitus induces hypercholesterolemia and a markedly increased predisposition to atherosclerosis.

Elevated plasma homocysteine[10]

Recent clinical and epidemiologic studies have shown a more general relationship between total serum homocysteine levels and coronary artery disease, peripheral vascular disease, stroke or venous thrombosis.

Factors affecting hemostasis and thrombosis

Epidemiologic evidence indicates that several markers of hemostatic and thrombotic function and inflammation are potent predictors of risk for major atherosclerotic events, including MI and stroke. Such markers include elevated plasminogen activator inhibitor, plasma fibrinogen and C-reactive protein (CRP).

Infection

Chlamydia pneumonia has been demonstrated in atherosclerotic plaques. This infectious organism incites a chronic inflammatory process that contributes to atheroma formation.

Lipoprotein (a)

It is an altered form of LDL-C that contains apo-B100 portion of the LDL-C linked to apo-A. It is a large glycoprotein molecule with a high degree of structural homology to plasminogen. Epidemiologic studies have shown a correlation between increased blood levels of Lipoprotien(a) and coronary and cerebrovascular disease, independent of the level of total cholesterol or LDL-C.

Stressful life style

It is type-A behaviour pattern, characterized by aggressiveness, competitive drive, ambitiousness and a sense of urgency is associated with enhanced risk of IHD.

Physical inactivity

Lack of exercise is associated with the risk of developing atherosclerosis and its complications.

1.3.2 Coronary Artery Disease (CAD)

Narrowing of the arteries that supply blood to the myocardium, and results in limiting blood flow and insufficient amounts of oxygen to meet the needs of the heart. The narrowing may progress to the extent that the heart muscle would sustain damage due to lack of blood supply.

1.3.3 Myocardial Infarction (MI)

MI is a condition which occurs when blood and oxygen supplies are partially or completely blocked from flowing in one or more cardiac arteries, resulting in damage or death of heart cells. The blockage is usually due to the formation of a clot in an artery. This condition is commonly known as heart attack. The occlusion may be due to ruptured atherosclerotic plaque. If the restricted flow of blood through the arteries and the resulting limited supply of oxygen are left untreated for a period of time, the blockage can cause damage or death of the myocardium cells.

1.3.4 Angina Pectoris

Currently, termed angina, this condition is not a disease but a symptom of an underlying heart condition. It is characterized by chest pain, discomfort or a squeezing pressure. The pain may also be felt in the shoulders, arms, neck and back. Angina occurs as a result of a reduction or a lack of blood supply to a part or the entire heart muscle, as well as impairment of waste removal. Poor blood circulation is usually due to CHD when partial or complete obstruction of the coronary arteries is present. Angina attacks may be due to spasm of the arteries. Angina may be a symptom of coronary microvascular disease (MVD), a condition that affects the heart's smallest arteries.

1.3.5 Stroke or Cerebrovascular Accident (CVA)

Stroke occurs when blood circulation in part of the brain is blocked or diminished. When blood supply, which carries oxygen, glucose, and other nutrients, is disrupted, brain cells die and become dysfunctional. Usually strokes occur due to blockage of an artery by a blood clot or a piece of atherosclerotic plaque that breaks loose in a small vessel within the brain.

1.4 MANAGEMENT OF HYPERLIPIDAEMIA:

The modalities of treatment used are:

1. Dietary modification
2. Increased physical activity
3. Elimination of associated risk factors. eg: smoking
4. Drug therapy.

1.4.1 Dietary Modification

Dietary intervention[11] is the primary treatment strategy, but drug therapy may often be added later to augment treatment. The main component of a "heart-healthy" diet is a food pattern that is low in saturated fat and dietary cholesterol and provides adequate energy to support growth and maintain an appropriate weight.

Decreased intakes of saturated fat

Reducing saturated fat is considered to have the most impact in lowering LDL. Sources include stick margarine, partially hydrogenated oils and fats, hydrogenated peanut butters, commercial bakery products, commercial fried food(e.g., French fries) and high fat animal products.

Intakes of trans-fatty acids

Trans-fatty acids are thought to increase LDL levels nearly as much as saturated fat and appear to lower HDL.

Decreased intakes of dietary cholesterol

Although individual responsiveness to dietary cholesterol varies, it is still considered important in LDL reduction. Diabetic subjects may be more sensitive to dietary cholesterol intake, which is only found in animal products.

Encourage a low to moderate total fat intake

Currently the specific type of fat consumed is emphasized over the total fat intake which was once considered to be the most important factor in lowering cholesterol. The AAP and AHA have placed upper and lower limits on fat intake toprevent nutrient deficiencies possible with very low fat diets and to avoid possible adverse effects of high carbohydrate diets upon HDL and TG’s.

Balance the fatty acid composition of diet

Polyunsaturated and monounsaturated fatty acids can lower LDL and are suggested as substitutes for saturated fats.

Encourage Omega-3-fatty acid consumption

Because of their association with lower TG and other cardio protectiveeffects, the AHA recommends at least 1 fatty fish meal or other source of Omega 3-fatty acids per week.

Increase dietary fiber intake

Soluble fiber can contribute to LDL reduction and is now a formal part of hyperlipidemia dietary recommendations. Fruits, vegetables, cereals, oats, wholegrains, and legumes are good sources of soluble fiber.

Encourage antioxidant food sources

Carotenoids vitamins C and E have been associated with lower CHD risk. Recommend antioxidant-rich foods such as whole grains, citrus fruits, melons, berries and dark orange/yellow or leafy green vegetables rather than supplements.

Reduce serum homocysteine levels:

Higher blood levels of homocysteine are associated with greater CHD risk. Adequate intakes of folate and vitamins B_6 and B_{12} as well as total fat restriction may keep homocysteine levels low. Food sources of these nutrients include fruits, dark green and leafy vegetables, fortified cereals, whole grains, lean meats and poultry.

Plant sterols are not recommended

Plant-sterol-containing foods are associated with lower LDL, but may decrease the absorption of fat-soluble vitamins.

1.4.2 Regular Physical Activity[12]

30 – 60 min of endurance (cardiovascular) activities (e.g., brisk walking, jogging, cycling) 4 – 7 days per week, benefits,

- Helps with weight loss / weight maintenance
- Helps control blood glucose
- Lowers the level of LDL and TG's
- Increases HDL-C
- Helps lower blood pressure
- Decreases the risk for heart disease
- Helps improve sleep quality
- Increases circulation in the body
- Improves wellbeing.

1.4.3 Maintain Ideal Body Weight

Advice subjects with dyslipidemia who are overweight (Body mass index [BMI] > 25) or have a waist circumference > 90 cm (women) or > 100 cm (Men) to reduce their weight.

Encourage subjects to attain and maintain a healthy body weight (BMI of 20 – 25)

1.4.4 Consume Alcohol in Moderation

Patients who choose to drink should limit their alcohol consumption to 2 or fewer standard drinks per day. Advise patient with elevated triglyceride levels to decrease or eliminate alcohol consumption.

1.4.5 Stop Smoking

Advise patients who smoke, to quit and encourage young people not to smoke. Provide subjects who are unable to quit on their own with information on smoking cessation programs, nicotine replacement therapy and drug therapy where indicated.

1.4.6 Drug Therapy

1.4.6.1 Statins

Statins are the most common medications used in the treatment of Hyperlipidaemia. Last year more than 200 million patients were treated with Statins. They are also referred to as HMG-CoA Reductase Inhibitors because of their mechanism of action. They are well tolerated and are effective in lowering LDL. Additionally, they have the highest level of patient compliance due to their tolerable adverse effects. Statins are useful for high-risk subjects such as those with CHD and diabetes.

Statins act by interfering in the biosynthesis of cholesterol in the liver. This is achieved by inhibiting the enzyme HMG-CoA reductase and reducing the rate by which it is able to produce mevalonate which is required in the biosynthesis. In addition to their cholesterol-lowering effect, Statins have anti-atherosclerotic activity. They enhance the stability of atherosclerotic plaques and exert pleiotropic effect (endothelial function, inflammation, coagulation and plaque vulnerability).

Currently used statins are lovastatin, pravastatin, Simvastatin, fluvastin, atorvastatin, rosuvastatin and pitavastatin. Comparison of efficacy of these drugs revealed that atorvastatin resulted in reduction of LDL of 42%; lovastatin and Simvastatin made reductions of 36% each. Results of triglycerides reduction were atorvastatin 19%, Simvastatin 13% and lovastatin 12%. Serum HDL level increased by 5% - 6% with all Statins.

In 2011, the FDA announced safety label changes for Simvastatin which include limiting the use of the highest approved dose of 80 mg due to increased risk of myopathy, particularly during the first 20 months. The most frequently encountered side effects include abdominal distress, constipation, flatulence, nausea, heartburn and headache. Myalgia and/or muscle weakness are rarely reported. The usual initial dose for adults is 20 mg daily.

A dosage of 20 mg daily is initiated and this may be increased at intervals of no less than 4 weeks until a maximum dosage of 80 mg is reached.

Lovastatin, which occurs naturally and is found in food such as oyster mushrooms, was the first Statin to be approved by the FDA. It is partially absorbed from the GI tract and undergoes first pass extraction. Food appears to enhance the rate of absorption after oral administration. The side effects, which include abdominal pain, cramps, and dyspepsia, are usually mild and transient. As with all Statins, dosage of lovastatin varies from one person to another and should be determined in accordance with the requirement and response of the patient. The usual maintenance dose is 10 – 80 mg daily given in a single or divided dose.

The usual maintenance dose for Pravastatin is 10 – 40 mg daily. The drug can be taken with meals, as food does not appear to affect its activity. Pravastatin side effects include nausea, vomiting, diarrhea, abdominal cramps, flatulence, headache, constipation and muscular pain.

Fluvastatin possesses a low incidence of side effects that are usually well tolerated. The most common ones are abdominal discomfort, headache, back pain and rash.

Atorvastatin is used to reduce LDL and triglycerides concentration. It is usually taken once a day with or without food. It is contraindicated in pregnancy and its intake by breast-feeding mothers is not recommended.

Side effects include headache, weakness, insomnia, rash, abdominal discomfort, constipation and diarrhea. Adult dose of atorvastatin should be adjusted within 2-4 weeks after the initial dose of 10 mg daily. The maintenance daily adult dose is 10 – 80 mg daily.

As with all Statins, there is a concern of development of rhabdomyolysis following the use of rosuvastatin. However, the FDA indicated that the risk of this condition is greater with rosuvastatin than with other marketed Statins. The FDA indicated that the risk of myopathy during rosuvastatin therapy may be increased in asian-americans. Physicians should start asian subjects at the lower dose level. The starting dose for most adults is 5 mg once daily and the maximum dose is 40 mg per day.

1.4.6.2 Fibric Acid Derivatives

Gemfibrozil (Lopid®) and Fenofibrate (Tricor®) have minimal effect on lowering LDL blood level, but are effective in reducing plasma triglyceride content by increasing fatty acid oxidation in the liver, thereby reducing secretion of VLDL. Additionally, they can increase HDL levels. These drugs can be used in combination with niacin or bile acid sequestrants. The most encountered adverse effects of fibric acid derivatives include rash and gastrointestinal disturbances. Statin fibrate combination therapy resulted in a 35% - 42% decrease in LDL, a 48% - 57% decrease in triglycerides and an increase of 14% - 17% in HDL.

1.4.6.3 Bile Acid Sequestrants

The bile acid binding resins, cholestyramine, colestipol and colesevelam, combine with bile acids present in the intestine to form an insoluble complex. This leads to an increase in LDL receptors and a reduction in plasma LDL. These medications may be used alone or in combination with Statins. Because the bile acid sequestrants are not absorbed from the GI tract, they do not possess systemic adverse effects. However, they are associated with GI tract disturbances such as constipation, nausea, flatulence and indigestion.

1.4.6.4 Cholestyramine

This drug is taken orally as a suspension prepared from a powder. Caution should be exercised not to take the powder in the dry form as it may cause esophageal irritation or blockage. The usual initial adult dose is 3 gm, 3 times daily before meals. The maintenance dose is 4 gm, 3 times daily before meals and at bedtime.

1.4.6.5 Colestipol Hydrochloride

This is a high molecular weight basic anion-exchange resin. The mechanism of action, adverse effects, and mode of administration are similar to those of cholestyramine. It is dispensed in tablet and granular forms. The usual adult dose is 1 to 16 gm daily.

1.4.6.6 Ezetimibe

Ezetimibe was approved by the FDA in 2002 for subjects with low risk of CHD and inability to tolerate Statins. Its mechanism of action differs from that of Statins. It is considered a selective cholesterol absorption inhibitor. It acts by blocking the absorption of dietary and biliary cholesterol. Recommended dose is 10 mg daily. Side effects include GI disturbances, headache, fatigue, myalgia, rash, and very rarely, myopathy.

1.4.6.7 Plant Sterols

These are capable of lowering LDL by about 10%. Their mechanism of action involves blocking cholesterol absorption from the intestines. Plant sterols are available as nonprescription drugs and should not be recommended as a primary therapy.

1.4.6.8 Niacin

Niacin is capable of reducing LDL (15% - 25%), VLDL (25% - 35%) and triglycerides concentration and at the same time results in elevation of HDL (15% - 25%). The mechanism of action of niacin is not fully known, but it has been postulated that niacin can partially inhibit free fatty acid release from adipose tissue and reduce the rate of synthesis of VLDL. The main adverse effects of niacin include uncomfortable and potentially dose limiting flushing of the skin, itching, skin rash, GI disturbances, hepatotoxicity and an increase in blood sugar and uric acid. The usual adult maintenance dose is 1 to 2 gm three times daily after meals.

1.4.6.9 Other Drugs

Omega 3 fatty acids obtained from fish liver oil are long chain highly polyunsaturated, principally eicosapentaenoate and docosahexaenoate reduces the triglycerides and LDL level by reducing the amount of cholesteryl esters in nascent VLDL and increases HDL level by the reducing the concentration of free fatty acids in plasma causing reduced net flux of cholesteryl esters from HDL to LDL and VLDL via reduced activity of the cholesteryl ester transfer protein.

Therapeutic lifestyle changes (TLC) as well as initiation of the intake of cholesterol lowering drugs are helpful in reducing LDL blood levels and at the same time may increase HDL blood levels, thereby reducing the risk of the occurrence of cardiovascular disease and cerebrovascular accidents. Combination therapy of Statins, niacin, fibrate or bile acid sequestrants are helpful in lowering LDL and triglyceride blood levels.

1.4.6.10 Combining lipid-modifying drugs[13]

a. **Bile acid binding resin plus Statins**: Provides significant additional LDL-C–lowering efficacy. Colestipol with lovastatin has been shown to decrease atherosclerosis progression and increase regression.

b. **Statins plus Ezetimibe**[14]: Through the dual inhibition of cholesterol absorption and synthesis, ezetimibe plus simvastatin allowed more patients to reach LDL cholesterol< 100 mg/dl at a lower simvastatin dose and with fewer dose titrations than simvastatin monotherapy.

c. **Niacin plus Statins:** Shown to induce regression of CIMT in subjects with CHD or CHD risk equivalents.

d. **Fenofibrate and Statin**: Effective in reducing[15] triglyceride levels and raising HDL-C levels in subjects with mixed dyslipidaemia. Niacin or Fibrate in combination with Statins be aware of the possible adverse effects of each agent.

e. **Bile acid binding resin plus niacin**[16]**:** Effective in familial hypercholesterolemia (type-IIa) and also in familial combined hyperlipidemia (Type-IIb). The resin has acid neutralising action which reduces gastric irritation caused by niacin. There is quantitative evidence of reversal of CHD.

f. **Omega-3 fatty acids and Statins**[17]: Omega-3 fatty acids, which contain marine-derived long-chain fatty acids Eicosapentaenoic acid (EPA) and Docosahexaenoic acids (DHA), have been shown to decrease circulating triglyceride concentrations by 20% to 50% through the reduction of hepatic very low-density lipoprotein (VLDL) synthesis and secretion.

The blood lipids of subjects with mixed dyslipidaemia have been shown to respond favourably to omega-3/statin combinations and have multiple lipid-lowering benefits, which included decreases in VLDL cholesterol, triglycerides and total cholesterol, respectively.

Furthermore, multidrug combinations using Omega-3 fatty acid supplementation has also been suggested as a potential approach in the treatment of dyslipidaemia. This "poly-portfolio" secondary prevention strategy including high-dose Statin, 3 antihypertensive medications, aspirin, and Omega-3 fish oil, was estimated to reduce major heart events substantially in subjects with cardiovascular disease, although this effect has yet to be demonstrated.

1.4.7 Rationale for Combination Therapy[18]

Novel approaches to lipid management of two or more hypolipidaemic drugs may be required to meet target LDL, TC and TG blood levels. Combinations of drugs that act by different mechanisms can provide additive effects in LDL, TC and TG reduction. When lipid-modifying agents are co-administered, both the exogenous and endogenous pathways of cholesterol metabolism are affected for dual activity and broader lipid control.

In Combination therapy, the rationale of using lower dosages of more than one drug to avoid the side effects from higher doses of either drug, or a second medication is added to a moderate or high amount of drug to optimize the ability to achieve goals of therapy. A growing literature suggests that combination therapy in those with severe or complex forms of hyperlipidemia can improve the lipid profile and reduce clinical endpoints.

Despite the low-density lipoprotein cholesterol lowering efficacy of Statins, many subjects, especially those at the highest risk for coronary heart disease (CHD), do not reach LDL-C goals. Although there are many reasons for this observation, one possible solution is to treat hyperlipidemic subjects with a combination of two lipid-altering agents that have different mechanisms of action.

A distressing result of the Lipid Treatment Assessment Project study was the finding that those at highest risk of coronary heart disease (CHD) were least likely to be treated to goal. Many reasons have been offered, but one attractive solution is combination therapy of lipid-lowering drugs.

Resins certainly augment LDL-C lowering, but gastrointestinal side-effects are considerable. In clinical trials where resins were used, the compliance rates were diminished by increased gastrointestinal side effects including constipation, bloating, flatulence, and abdominal distress. Niacin can raise blood sugar and uric acid and cause abnormal liver profiles. With care it can be used successfully in those who have impaired fasting glucose or glucose tolerance. While extended- and sustained-release forms can diminish flushing, niacin still requires careful monitoring.

Fibrates are generally well tolerated, but there is a concern over myositis and rhabdomyolysis when combined with Statins. Several trials have illustrated the potential utility of combining resins with Statins or niacin, and niacin or fibrates with Statins. However, many subjects poorly tolerate resins and niacin, and fibrates combined with Statins may increase the risk of myopathy.

An 'ideal' second drug[19] to add to Statins or niacin would:

(a) Lower LDL-C by 15–18%;

(b) Not potentiate the liver or muscle toxicities associated with high dose Statins and

(c) Cause few side effects or drug-drug interactions that would compromise compliance. Ezetimibe is a new cholesterol absorption inhibitor that exhibits all of these characteristics.

Ezetimibe inhibits[20] the intestinal absorption of dietary and biliary cholesterol without interfering with the absorption of fat-soluble vitamins. When added to Statins, ezetimibe produces significant reductions in LDL-C and triglycerides beyond reductions seen with a Statin alone. Moreover, ezetimibe has a favorable safety and tolerability profile without any clinically important drug interactions. Thus, the combination of statins with ezetimibe offers a promising new approach to lowering LDL-C.

A second drug[21] should have a complementary lowering effect on LDL-C by at least 15–18% with a single dose. Because of the nonlinear log dose–response curve of statins, when the initial Statin dose is doubled, there is only an additional 6% lowering of LDL-C. Thus, a medication that would provide an 18% lowering of LDL-C would save three doublings of the initial Statin dose to obtain the same effect on LDL-C. In addition, it should improve other aspects of the lipid panel and not aggravate any aspects of the risk factor profile.

A second drug should not increase the potential liver or muscle toxicity of Statins. Fortunately, both effects are infrequent, but the recent adverse experience with cerivastatin has caused physicians to be hesitant about using higher doses of Statins as they wish to avoid either the increased liver transaminase elevations and/or muscle problems more frequently seen at the higher Statin dosages. In the Expanded Clinical Evaluation of Lovastatin (EXCEL) trial, the 80 mg dose of lovastatin caused more than twice the liver and/or muscle problems than did the 40 mg dose.

Tolerability factors have a strong bearing on compliance. A second drug should have high tolerability with few side effects to limit compliance. It ideally would have few drug–drug interactions. Convenience (once-daily dosing) would be an important factor to aid compliance. Finally, cost considerations would be important in any multidrug regimen.

Possible 'second' drugs that might fit the criteria for the 'ideal' second drug to combine with low-dose Statin therapy is ezetimibe, a novel intestinal cholesterol absorption inhibitor that blocks dietary and biliary cholesterol absorption at the intestinal brush border, appears to circumvent some of the problems seen with the addition of resins and plant stanol/sterol esters to Statins.

For those on a lower-dose statin who add ezetimibe, the combination may end up being safer and more tolerable than the highest doses of a potent Statin. Longer-term phase III data on ezetimibe suggest that this is a safe drug and may be an alternative to those who cannot take Statin therapy as well as being useful in combination. Moreover, ezetimibe is very tolerable. It need not be taken with food and can be taken either in the morning or evening as a single10 mg tablet.

Although Statins like Simvastatin monotherapy reduces TC, LDL-C and apo-B, it achieved only limited TG reduction. Fibrate therapy lowers the elevated TC levels, but limited LDL reduction. Combination of fibrates and statins can lead to increased risk of myopathy. Monotherapy with Omega-3 fatty acids even though efficiently reduces TG, LDL reduction is inconsistent.

Combinations like Omega-3 fatty acids (4 g) with Simvastatin (20 mg) or with Ezetimibe (10 mg) reduced elevated total and LDL cholesterol and triglycerides in mixed dyslipidemic condition without adverse events. Simvastatin and Omega-3 fatty acids combination is also usefull in subjects with low level of HDL.

This study tested the hypothesis that triple combination (20 mg Simvastatin, 10 mg Ezetimibe and 4 g of Omega-3 fatty acids) would reduce elevated total cholesterol, LDL cholesterol, triglycerides and elevate the level of HDL in mixed dyslipidemic subjects.

Optimal Plasma Lipid Levels [22]

Total cholesterol:

< 200 mg/dl Desirable

200-239 mg/dl Borderline high

≥ 240 mg/dl High

HDL-C:

< 40 mg/dl Low (consider <50 mg/dl as low for women)

> 60 mg/dl High

LDL-C:

< 70 mg/dl Optimal for very high risk

< 100 mg/dl Optimal

100-129 mg/dl near optimal 130-

159 mg/dl Borderline high 160-189

mg/dl High

≥ 190 mg/dl Very high.

Triglycerides:

< 150 mg/dl Normal

150-199 mg/dl Borderline high

200-499 mg/dl High

≥ 500 mg/dl Very high.

1.5 DRUG PROFILE

1.5.1 Simvastatin

Simvastatin[23] is a lipid-lowering agent that is derived synthetically from a fermentation product of Aspergillus terreus. After oral ingestion, Simvastatin, which is an inactive lactone, is hydrolyzed to the corresponding β-hydroxyacid form. This is an inhibitor of 3-hydroxy-3-methylglutaryl-coenzyme A (HMG-CoA) reductase. This enzyme catalyzes the conversion of HMG-CoA to mevalonate, which is an early and rate-limiting step in the biosynthesis of cholesterol.

Chemistry

Simvastatin is butanoic acid, 2,2-dimethyl-,1,2,3,7,8,8a-hexahydro - 3,7- dimethyl- 8- [2-(tetrahydro-4hydroxy-6-oxo- 2H- pyran- 2- yl)- ethyl]-1-naphthalenyl ester, [1S-[1α,3α,7β,8β(2S*,4S*),-8aβ]]. The empirical formula of Simvastatin is $C_{25}H_{38}O_5$ and its molecular weight is 418.57. Its structural formula is:

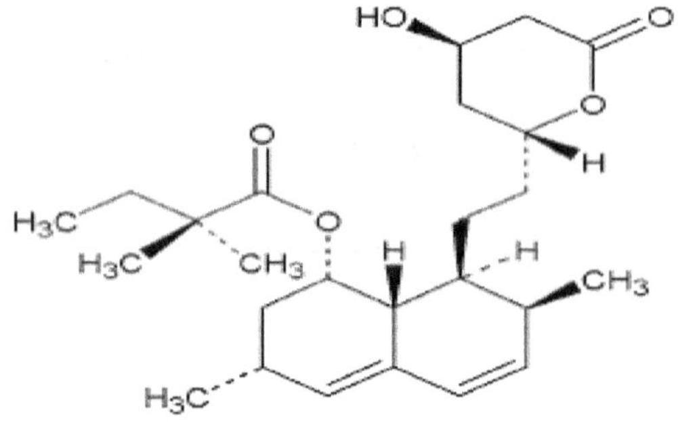

Simvastatin is a white to off-white, nonhygroscopic, crystalline powder that is practically insoluble in water, and freely soluble in chloroform, methanol and ethanol.

Simvastatin tablets for oral administration contain either 5 mg, 10 mg, 20 mg, 40 mg or 80 mg of Simvastatin and the following inactive ingredients: ascorbic acid, citric acid, hydroxypropyl cellulose, hypromellose, iron oxides, lactose, magnesium stearate, microcrystalline cellulose, starch, talc, and titanium dioxide. Butylated hydroxyanisole is added as a preservative.

Mechanism of Action

Simvastatin is a prodrug and is hydrolyzed to its active β-hydroxyacid form, Simvastatin acid, after administration. Simvastatin is a specific inhibitor of 3-hydroxy-3-methylglutaryl-coenzyme A (HMG-CoA) reductase, the enzyme that catalyzes the conversion of HMG-CoA to mevalonate, an early and rate limiting step in the biosynthetic pathway for cholesterol. In addition, Simvastatin reduces VLDL and TG and increases HDL-C.

Pharmacodynamics

Epidemiological studies have demonstrated that elevated levels of total-C, LDL-C, as well as decreased levels of HDL-C are associated with the development of atherosclerosis and increased cardiovascular risk. Lowering LDL-C decreases this risk. However, the independent effect of raising HDL-C or lowering TG on the risk of coronary and cardiovascular morbidity and mortality has not been determined.

Pharmacokinetics

Simvastatin is a lactone that is readily hydrolyzed in vivo to the corresponding β-hydroxyacid, a potent inhibitor of HMG-CoA reductase. Inhibition of HMG-CoA reductase is the basis for an assay in pharmacokinetic studies of the β-hydroxyacid metabolites (active inhibitors) and, following base hydrolysis, active plus latent inhibitors (total inhibitors) in plasma following administration of Simvastatin.

Following an oral dose of 14C-labeled Simvastatin in man, 13% of the dose was excreted in urine and 60% in feces. Plasma concentrations of total radioactivity (Simvastatin plus 14C-metabolites) peaked at 4 hours and 40declined rapidly to about 10% of peak by 12 hours postdose. Since Simvastatin undergoes extensive first-pass extraction in the liver, the availability of the drug to the general circulation is low (< 5%).

Both Simvastatin and its β-hydroxyacid metabolite are highly bound (approximately 95%) to human plasma proteins. Rat studies indicate that when radiolabeled Simvastatin was administered, Simvastatin derived radioactivity crossed the blood-brain barrier.

The major active metabolites of Simvastatin present in human plasma are the β-hydroxyacid of Simvastatin and its 6'-hydroxy, 6'-hydroxymethyl, and 6'-exomethylene derivatives. Peak plasma concentrations of both active and total inhibitors were attained within 1.3 to 2.4 hours postdose. While the recommended therapeutic dose range is 5 to 40 mg/day, there was no substantial deviation from linearity of AUC of inhibitors in the general circulation with an increase in dose to as high as 120 mg. Relative to the fasting state, the plasma profile of inhibitors was not affected when Simvastatin was administered immediately before an American Heart Association recommended low-fat meal.

Indications

Therapy with lipid-altering agents should be only one component of multiple risk factor intervention in individuals at significantly increased risk for atherosclerotic vascular disease due to hypercholesterolemia. Drug therapy is indicated as an adjunct to diet when the response to a diet restricted in saturated fat and cholesterol and other nonpharmacologic measures alone has been inadequate. In subjects with coronary heart disease (CHD) or at high risk of CHD, Simvastatin can be started simultaneously with diet.

Dosage and Administration

The usual dosage range is 5 to 40 mg/day. In subjects with CHD or at high risk of CHD, Simvastatin can be started simultaneously with diet. The recommended usual starting dose is 10 or 20 mg once a day in the evening. For subjects at high risk for a CHD event due to existing CHD, diabetes, peripheral vessel disease, history of stroke or other cerebrovascular disease, the recommended starting dose is 40 mg/day. Lipid determinations should be performed after 4 weeks of therapy and periodically thereafter.

Precautions

Simvastatin occasionally causes myopathy manifested as muscle pain, tenderness or weakness with creatine kinase (CK) above ten times the upper limit of normal (ULN). Myopathy sometimes takes the form of rhabdomyolysis with or without acute renal failure secondary to myoglobinuria, and rare fatalities have occurred. The risk of myopathy is increased by high levels of Statin activity in plasma. Predisposing factors for myopathy include advanced age (≥ 65 years), female gender, uncontrolled hypothyroidism, and renal impairment.

1.5.2 Ezetimibe

Ezetimibe[24] is in a class of lipid-lowering compounds that selectively inhibits the intestinal absorption of cholesterol and related phytosterols. The chemical name of Ezetimibe is 1-(4-fluorophenyl)-3(R)-[3-(4-fluorophenyl)-3(S)-hydroxypropyl]-4(S)-(4-hydroxyphenyl)-2-azetidinone. The empirical formula is $C_{24}H_{21}F_2NO_3$. Its molecular weight is 409.4 and its structural formula is:

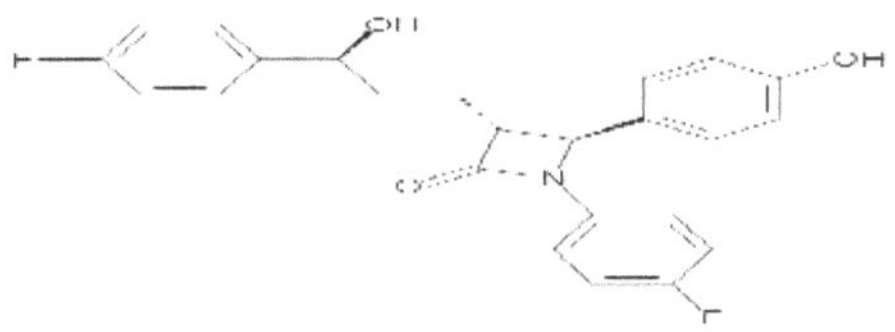

Ezetimibe is a white, crystalline powder that is freely to very soluble in ethanol, methanol, and acetone and practically insoluble in water. Ezetimibe has a melting point of about 163°C and is stable at ambient temperature. Ezetimibe is available as a tablet for oral administration containing 10 mg of Ezetimibe and the following inactive ingredients:

croscarmellose sodium NF, lactose monohydrate NF, magnesium stearate NF, microcrystalline cellulose NF, povidone USP, and sodium lauryl sulfate NF.

Mechanism of Action

Ezetimibe reduces blood cholesterol by inhibiting the absorption of cholesterol by the small intestine. In a 2-week clinical study in 18 hypercholesterolemic subjects, Ezetimibe inhibited intestinal cholesterol absorption by 54%, compared with placebo. Ezetimibe had no clinically meaningful effect on the plasma concentrations of the fat-soluble vitamins A, D, and E , and did not impair adrenocortical steroid hormone production .

The cholesterol content of the liver is derived predominantly from three sources. The liver can synthesize cholesterol, take up cholesterol from the blood from circulating lipoproteins, or take up cholesterol absorbed by the small intestine. Intestinal cholesterol is derived primarily from cholesterol secreted in the blle and from dietary cholesterol.

Ezetimibe has a mechanism of action that differs from those of other classes of cholesterol-reducing compounds (Statins, bile acid sequestrants [resins], fibric acid derivatives, and plant stanols). The molecular target of Ezetimibe has been shown to be the sterol transporter, Niemann-Pick C1-Like 1 (NPC1L1), which is involved in the intestinal uptake of cholesterol and phytosterols.

Ezetimibe does not inhibit cholesterol synthesis in the liver, or increase bile acid excretion. Instead, Ezetimibe localizes at the brush border of the small intestine and inhibits the absorption of cholesterol, leading to a decrease in the delivery of intestinal cholesterol to the liver. This causes a reduction of hepatic cholesterol stores and an increase in clearance of cholesterol from the blood; this distinct mechanism is complementary to that of Statins and of fenofibrate.

Pharmacokinetics

After oral administration, Ezetimibe is absorbed and extensively conjugated to a pharmacologically active phenolic glucuronide (Ezetimibe-glucuronide). After a single 10-mg dose of Ezetimibe to fasted adults, mean Ezetimibe peak plasma concentrations (Cmax) of 3.4 to 5.5 ng/mL were attained within 4 to 12 hours (Tmax). Ezetimibe-glucuronide mean Cmax values of 45 to 71 ng/mL were achieved between 1 and 2 hours (Tmax).

There was no substantial deviation from dose proportionality between 5 and 20 mg. The absolute bioavailability of Ezetimibe cannot be determined, as the compound is virtually insoluble in aqueous media suitable for injection.

Indications

Therapy with lipid-altering agents should be only one component of multiple risk factor intervention in individuals at significantly increased risk for atherosclerotic vascular disease due to hypercholesterolemia. Drug therapy is indicated as an adjunct to diet when the response to a diet restricted in saturated fat and cholesterol and other nonpharmacologic measures alone has been inadequate.

Dosage and Administration

The recommended dose of Ezetimibe is 10 mg once daily.

Side Effects

The most common adverse reactions in the group of subjects treated with Ezetimibe are:

Arthralgia (0.3%)

Dizziness (0.2%)

Gamma-glutamyltransferase increased (0.2%)

Overdose

In clinical studies, administration of Ezetimibe, 50 mg/day to 15 healthy subjects for up to 14 days, 40 mg/day to 18 subjects with primary hyperlipidemia for up to 56 days, and 40 mg/day to 27 subjects with homozygous sitosterolemia for 26 weeks was generally well tolerated. One female patient with homozygous sitosterolemia took an accidental overdose of Ezetimibe 120 mg/day for 28 days with no reported clinical or laboratory adverse events.

In the event of an overdose, symptomatic and supportive measures should be employed.

1.5.3 Omega 3 Fatty Acids

Omega 3 fatty acids, a lipid-regulating agent, are supplied as a liquid-filled gel capsule for oral administration. Each 1-gram capsule of Omega 3 fatty acids contains at least 900 mg of the ethyl esters of Omega-3 fatty acids sourced from fish oils. These are predominantly a combination of ethyl esters of eicosapentaenoic acid (EPA - approximately 465 mg) and docosahexaenoic acid (DHA - approximately 375 mg).

The empirical formula of EPA ethyl ester is $C_{22}H_{34}O_2$, and the molecular weight of EPA ethyl ester is 330.51.

The structural formula of EPA ethyl ester is:

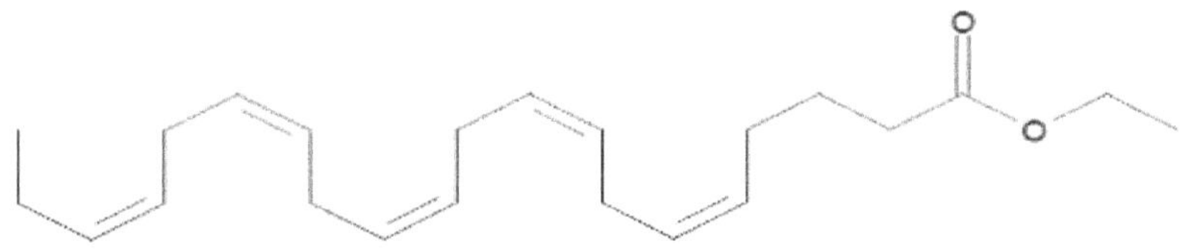

The empirical formula of DHA ethyl ester is $C_{24}H_{36}O_2$, and the molecular weight of DHA ethyl ester is 356.55.

Structural formula of DHA ethyl ester is:

Omega 3 fatty acids capsules also contain the following inactive ingredients: 4 mg a-tocopherol (in a carrier of soybean oil), and gelatin, glycerol, and purified water (components of the capsule shell).

Mechanism of Action

Potential mechanisms of action include inhibition of acyl-CoA:1,2-diacylglycerol acyltransferase, increased mitochondrial and peroxisomal β-oxidation in the liver, decreased lipogenesis in the liver, and increased plasma lipoprotein lipase activity. Omega 3 fatty acids may reduce the synthesis of triglycerides in the liver because EPA and DHA are poor substrates for the enzymes responsible for TG synthesis, and EPA and DHA inhibit esterification of other fatty acids.

Pharmacokinetics

In healthy volunteers and in subjects with hypertriglyceridemia, EPA and DHA were absorbed when administered as ethyl esters orally. Omega-3-acids administered as ethyl esters (Omega 3 fatty acids) induced significant, dose-dependent increases in serum phospholipid EPA content, though increases in DHA content were less marked and not dose-dependent when administered as ethyl esters.

Indications

Omega 3 fatty acids (Omega-3-acid ethyl esters) is indicated as an adjunct to diet to reduce triglyceride (TG) levels in adult subjects with severe (>500 mg/dL) hypertriglyceridemia (HTG).

Usage Considerations

Patients should be placed on an appropriate lipid-lowering diet before receiving Omega 3 fatty acids and should continue this diet during treatment with Omega 3 fatty acids.

Laboratory studies should be done to ascertain that the lipid levels are consistently abnormal before instituting therapy with Omega 3 fatty acids. Every attempt should be made to control serum lipids with appropriate diet, exercise, weight loss in obese subjects, and control of any medical problems such as diabetes mellitus and hypothyroidism that are contributing to the lipid abnormalities. Medications known to exacerbate hypertriglyceridemia (such as beta blockers, thiazides, estrogen) should be discontinued or changed if possible prior to consideration of triglyceride-lowering drug therapy.

Dosage and Administration

Assess triglyceride levels carefully before initiating therapy. Identify other causes (e.g., diabetes mellitus, hypothyroidism, medications) of high triglyceride levels and manage as appropriate.

Patients should be placed on an appropriate lipid-lowering diet before receiving Omega 3 fatty acids, and should continue this diet during treatment with Omega 3 fatty acids. In clinical studies, Omega 3 fatty acids was administered with meals. The daily dose of Omega 3 fatty acids is 4 grams per day.

Despite the low-density lipoprotein cholesterol (LDL-C) lowering efficacy of Statins, many subjects, especially those at the highest risk for coronary heart disease (CHD), do not reach LDL-C goals. Although there are many reasons for this observation, one possible solution is to treat hyperlipidemic subjects with a combination of two lipid-altering agents that have different mechanisms of action.

2. REVIEW OF LITERATURE

Todd C R ***et al***., reviewed[25] that expense, high drug dose, and low compliance to strict dietary therapies are current issues surrounding modern drug- and diet- based lipid lowering approaches. Further more, variable patient outcomes and suboptimal response to both drug and diet therapies are increasingly evident. Therefore, the question arises as to whether more emphasis should be placed on combination diet/drug therapies to reduce cholesterol levels in patients who respond suboptimally to diet and drug monotherapies. Although considerable research has explored multidrug combination therapies, combination drug/diet therapies receive less attention. However combined drug/diet approaches may reduce the number of drug prescriptions, the progressive increase in "optimal" drug dosage and costs associated with pharmaceutical disease management. Future research priorities in drug/diet therapeutic approaches should not only emphasize the discovery of novel combinations but also should address potential safety issues prior to wide-scale acceptance in clinical practice.

Jeffrey B R ***et al***., studied[26] consistency of effect of Ezetimibe/Simvastatin compared with intensified lipid-lowering treatment strategies in obese and non-obese diabetic subjects, In obese subjects (n=466), percent changes in LDL-C and most other lipids were greater with Ezetimibe/Simvastatin vs doubling the baseline statin dose or switching to rosuvastatin. In non-obese subjects (n=342) percent changes in LDL -C, total cholesterol, non-HDL-Cholesterol, apo B and apo A-1 were greater with ezetimibe/simvastatin vs doubling the baseline Statin dose or switching to rosuvastatin.

Hayati M Y ***et al.***, determined[27] the effects of prescription Omega-3(n-3) fatty acid ethyl esters (omacor®) on blood pressure, plasma lipids, and inflammatory marker concentrations in patients awaiting carotid endarterectomy. In conclusion, the study found that omacor given at 2 g/day for an average of 21 days to subjects with advanced carotid atherosclerosis lowers triglycerides and soluble E-selectin concentrations, but has limited broad impact on the plasma lipid profile or on inflammatory markers.This may be because the duration of intervention was too short or the dose of n-3 fatty acids was too low.

Charles J G ***et al.,*** studied[28] Omega 3 fatty acids from 4 to 8 to 12 gms/day in subjects with primary hypertriglyceridemia who had triglycerides levels >500mg/dl despite conventional triglyceride lowering therapy.Titration of Omega 3 fatty acids from 4 to 8 to 12 g/day safely offers an effective way to lower TG beyond conventional 4 g therapy.

Sang-Hyun K ***et al.,*** studied[29] a Prospective randomized comparison between Omega-3 fatty acid supplements plus Simvastatin versus Simvastatin alone in Korean subjects with mixed dyslipidemia: lipoprotein profiles and heart rate variability. After 6 weeks of drug treatment, triglyceride levels decreased by 41% in the combination treatment group and 13.9% in the Simvastatin monotherapy group.No significant changes in the heart rate variability (HRV) parameters were observed in either group.

Michel F ***et al.,*** reported[30] through post-hoc analysis, placebo treated subjects had the highest incidence of HDL-C reductions from baseline (45%) compared with subjects taking fenofibrate (14%), Ezetimibe+fenofibrate (9%) or Ezetimibe/ Simvastatin+fenofibrate (9%).

Leiter L A ***et al.,*** reported[31] the treatment of Ezetimibe/Simvastatin vs. Statin monotherapy provided significantly larger reductions in LDL-C, TC, TG, non-HDL-C, apo B and hs-CRP and significantly greater increases in HDL-C, with a similar safety profile in patients with and without diabetes.Reductions in LDL-C, TC, and non-HDL-C were larger in subjects with diabetes than in patients without diabetes.

Peter P T ***et al.,*** reported[32] an overview on how to best position lipid-lowering drugs when attempting to normalize serum lipid profiles and reduce risk for cardiovascular disease. Statins are widely accepted to be the agents of choice for reducing serum levels of LDL-C in both the primary and secondary prevention settings. Ezetimibe and bile acid sequestrants are both effective agents for reducing LDL-C, either used alone or in combination with Statins. The Statins, fibric acid derivatives (fibrates) and niacin raise high-density lipoprotein cholesterol to different extents depending upon genetic and metabolic background. Fibrates, niacin and Omega-3 fish oils are efficacious therapies for reducing serum triglycerides.Combinations of these drugs are frequently required for normalizing mixed forms of dyslipidaemia.

Gianluca B ***et al.,*** studied[33] a multicentre, randomized trial that Ezetimibe plus Simvastatin 10/20 mg provided a superior alternative for LDL-

C lowering vs doubling the dose of Simvastatin to 40mg in hyperlipidemic subjects with type2 diabetes mellitus and coronary heart disease.

Kater A-L A ***et al.,*** evaluated[34] whether the combination of Simvastatin and Ezetimibe results in a synergistic lowering effects on lipid and pro-inflammatory profiles in pre-diabetic subjects.They confirmed the synergistic lowering effects of a Simvastatin and Ezetimibe combination on LDL-cholesterol, apolipoprotein B, and triglycerides levels in subjects with early disturbances of glucose metabolism. They suggested an additive effect of this combination also on inflammatory status based on the reduction of C-reactive protein. Attenuation of pro-inflammatory conditions may be relevant in reducing cardiometabolic risk.

Thomas S ***et al.,*** studied[35] the changes in cholesterol absorption and cholesterol synthesis caused by Ezetimibe and/or Simvastatin in men. This was a randomized ,double blind,placebo-controlled ,four-period crossover study to evaluate the effects of coadministering 10mg Ezetimibe with 20 mg Simvastatin (Ezetimibe/Simvastatin) on cholesterol absorption and synthesis relative to either drug alone or placebo in 41 subjects.Each treatment period lasted 7 weeks. Ezetimibe, Simvastatin and Ezetimibe/Simvastatin decreased plasma LDL-cholesterol by 20, 38 and 55% respectively.

Sheridan M H ***et al.,*** reviewed[36] Omega-3 ethylester concentrate and its use in secondary prevention post-myocardial infarction and the treatment of hypertriglyceridaemia.They found out oral Omega-3 ethylester concentrate 4000 mg/day plus Simvastatin or atorvastatin reduced triglyceride, non-high-density lipoprotein cholesterol and/or very-low-density lipoprotein cholesterol levels to a significantly greater extent than placebo plus Simvastatin or atorvaStatin. Omega-3 ethylester concentrate was generally well tolerated both as secondary prevention post-myocardial infarction and in the treatment of hypertriglyceridaemia.

Winkler K ***et al.,*** evaluated[37] the effect of combination of fluvastatin/fenofibrate (80/200 mg) combared with combination of Simvastatin/Ezetimibe (20/10 mg) in subjects with metabolic syndrome and type2 diabetes. Simvastatin and Ezetimibe was more efficient in reducing total cholesterol and LDL-Cholesterol. However, in subjects with small dense low-density lipoprotein, fluvastatin/ fenofibrate was more efficient in reducing triglycerides and increasing LDL radius.

Heinz D ***et al.,*** reviewed[38] which classes of lipid lowering drugs for which subjects, with respect to the types of lipids or lipoproteins they predominantly affect. Statins inhibit the de-novo synthesis of cholesterol. There was abundant evidence that Statins lower the rate of cardiovascular events. This is not the case for the Ezetimibe, which strongly lowers LDL cholesterol. Its potential to decrease vascular risk remains to be proven. The lipid lowering dose of Omega-3 fattyacids was 3-4g/day. The typical candidate for Omega-3-fattyacids is a patient with severe hypertriglyceridemia in the primary prevention setting or in the secondary prevention setting where

Omega-3-fattyacids can be a part of a combination with Statins,or Statins and nicotinic acid ,or Statins and fibrates.

Derosa G ***et al.,*** evaluated[39] the prevalence of Statin-associated adverse events in diabetic and non-diabetic subjects affected by polygenic hypercholesterolemia or combined Hyperlipidaemia and the efficacy and tolerability of treatment with Ezetimibe/Simvastatin 10/10 mg/day on the same subjects experiencing the adverse events.The efficacy and adverse effect profile of the Ezetimibe and Simvastatin combination appear to be good for both diabetic and non-diabetic subjects ,and in both conditions.

Masuda D ***et al.,*** administered[40] Ezetimibe 10 mg/day in ten subjects with typeIIb hyperlipidaemia for two months and examined the lipid and lipoprotein profiles during fasting and after an oral fat loading test. Ezetimibe improved fasting lipoprotein profiles and postprandial hyperlipidaemia by suppressing intestinal chylomicrons production in subjects with type IIb hyperlipidaemia and such treatment may prove to be effective in reducing the atherosclerosis.

Gideon R H ***et al.,*** found out[41] lipid lowering therapy did not affect the postprandial drop in high density lipoprotein-cholesterol (HDL-C) plasma levels in obese men with metabolic syndrome by a randomized double blind crossover trial.

Anouk V D G ***et al.,*** evaluated[42] the efficacy and safety of long-term coadministration of Ezetimibe and Simvastatin in adolescents with heterozygous familial hypercholesterolemia. Coadministration of Ezetimibe with Simvastatin was safe, well tolerated, and provided higher LDL-C reduction compared with Simvastatin alone in adolescents with heterozygous familial hypercholesterolemia studied upto 53 weeks.

Constance C ***et all.,*** evaluated[43] the efficacy of switching from atorvastatin 10 mg to Ezetimibe/Simvastatin 10/20 mg, Ezetimibe/Simvastatin 10/40 mg or doubling the dose of atorvastatin from 10 to 20 mg in patients with type 2 diabetes. Ezetimibe/Simvastatin 10/20 mg and 10/40 mg provided greater lipid-altering efficacy than doubling the dose of atorvastatin from 10 to 20 mg and were well tolerated in patients with type 2 diabetes.

Teddy K ***et al.,*** reviewed[44] Ezetimibe's metabolism, pharmacokinetics and drug interactions. The recommended dose of Ezetimibe 10 mg/day can be administered in the morning or evening without regard to food. Overall, Ezetimibe has a favourable drug-drug interaction profile, as evidenced by the lack of clinically relevant interactions between ezetimibe and a variety of drugs commonly used in patients with hypercholesterolaemia.

Susan B ***et al.,*** studied[45] consistency in efficacy and safety of Ezetimibe coadministered with Statins for treatment of hypercholesterolemia in women and men. Compared with Statin monotherapy (lovastatin or pravastatin 10,20 or 40mg;Simvastatin or atorvastatin 10,20,40 or 80mg), Ezetimibe plus Statin demonstrated greater efficacy in reducing blood levels of LDL-C,apolipoprotein B, and triglycerides and raising high density lipoprotein cholesterol. Eztimibe plus Statin was equally efficacious in women and men. Ezetimibe plus Statin was well tolerated and had a favourable safety profile in both patient subgroups.

Marc E ***et al.,*** reviewed[46] medical regulating therapy, current evidence, ongoing trials and future developments. Elevated low density lipoprotein-cholesterol and reduced high density lipoprotein cholesterol were well recognised coronary heart disease (CHD) risk factors. Statins reduce LDL-C by inhibiting the rate limiting step in cholesterol biosynthesis and reduced CHD event rates in primary and secondary prevention trials. The magnitude of this effect is not fully accounted for by LDL-C reduction alone and may relate to effects on other lipid parameters such as HDL-C and apolipoproteins B and A-I, as well as additional anti-inflammatory effects.

Jennifer M ***et al.,*** reviewed[47] cytochrome P450 drug interactions within the HMG-CoA reductase inhibitor class, and their clinical relevance. Variables that affect concentration-effect relationship including changes in the plasma concentration, concomitant lipid-lowering therapy or host genetic factors that code for different forms or amounts of metabolising enzymes and drug receptors. Subjects taking a CYP3A4-metabolised agent like Simvastatin should not be started on a CYP3A4 inhibitor or inducer without close monitoring.

Nordoy A ***et al.,*** evaluated[48] the efficiency and the safety of treatment with Simvastatin and Omega-3 fatty acids in patients with combined hyperlipidaemia. Study showed that Omega-3 fatty acids have an additive effect in lowering triglycerides, total cholesterol and apolipoprotein E when added to the Statins. Omega-3 fatty acids also have a moderate antihypertensive effect confirmed in this study and were without deleterious effect on carbohydrate metabolism.

David W ***et al.,*** reported[49] cytochrome P450 (CYP) enzyme system plays an important part in the metabolism of the Statins, leading to clinically relevant interactions with other agents particularly cyclosporine, erythromycin, itraconazole, ketoconazole and HIV protease inhibitors, that are also metabolised by this enzyme system. The CYP3A family metabolises lovastatin, Simvastatin, atorvastatin and cerivastatin. Rhabdomyolysis had occurred following the coadministration of cyclosporine, a potent CYP3A4 and p-glycoprotein inhibitor. Pharmacodynamically, there is an increased risk of myopathy when Statins are coprescribed with fibrates or nicotinic acid.

Teddy K ***et al.,*** studied[50] pharmacodynamic interaction between the new selective cholesterol absorption inhibitor Ezetimibe 10 mg and simvastatin 20 mg. They evaluated the potential for pharmacodynamic and pharmacokinetic interaction between ezetimibe 0.25, 1, or 10 mg and Simvastatin 10 mg and a pharmacodynamic interaction between ezetimibe 10 mg and simvastatin 20 mg. Evaluation of the tolerance of the coadministration of Ezetimibe and Simvastatin was a secondary objective. The coadministration of Ezetimibe at doses upto 10 mg with Simvastatin 10 or 20 mg daily was well tolerated and caused a significant additive reduction in LDL-C compared with simvastatin alone.

3. NEED FOR THE STUDY

Despite the diversity of available LDL, TG and TC lowering therapies[51], a significant proportion of the hypercholesterolaemic population is not attaining the recommended target LDL, TG and TC levels. To achieve this goal, high Statin doses may be necessary, which increases its adverse effects. Simvastatin mono therapy may be insufficient for the desirable reduction in LDL levels, a combination of lipid-lowering agents has become frequent in clinical practice. In particular, Statin and Ezetimibe combination has been shown to be very effective in reducing total and LDL cholesterol levels.

Monotherapy[52] with one of the Statins does not always lower low density lipoprotein cholesterol and triglycerides or raise high density lipoprotein cholesterol to the required extent and it is sometimes necessary to combine their administration with other lipid regulating drugs. For example in severe familial hypercholesterolemia even maximal doses of Statins do not always lower LDL- C sufficiently and an anion exchange resin is often added. Also, in mixed hyperlipidaemia, statins monotherapy may fail to reduce triglycerides and raise HDL-C to the desired levels and it may be necessary to add either nicotinic acid or a fibrate to achieve these objectives. Combination therapy also improves the response of patients who are refractory to Statins.

In a number of small studies[53] the combination of Statins and Omega-3 fatty acids has been consistently shown to be an effective, safe, and well-tolerated treatment for combined dyslipidaemia. Patients with recent myocardial infarction may also benefit from this combination. When considering risks and benefits of adding a second agent to Statins for treatment of combined dyslipidaemia, Omega-3 fatty acids provide additional lipid improvements without requiring additional laboratory tests and do not increase risk for adverse muscle or liver effects.

Recently, as the target goals of triglycerides levels became tougher, a combination therapy of Omega-3 fatty acids and Statins would be a reasonable option for high-risk patients with combined or mixed dyslipidemia. Omega 3 fatty acids[54] lower the serum triglyceride levels. The HDL cholesterol concentration generally increases during Omega 3 therapy in patients with hypertriglyceridemia.

Thus, there is a continued search for effective, better-tolerated drugs or combinations of drugs for the treatment of patients with Hypercholesterolaemia to achieve the target goal of LDL, TC, TG and HDL levels.

The combined therapy of Ezetimibe, Simvastatin and Omega 3 fatty acids offers a new treatment strategy to attain better target levels of lipid that cannot be achieved with Simvastatin or Ezetimibe alone.

4. OBJECTIVES AND HYPOTHESES

It was hypothesized that more subjects will reach their LDL, HDL, TC and TG targets earlier with combined therapy of ezetimibe, simvastatin and omega 3 fatty acids with less tolerance and adverse events than those associated with high-dose Statin or Ezetimibe mono therapy. An extensive literature survey carried out by the author revealed that there are however, no reports of this combined therapy of these drugs in subjects with hyperlipidemia.

The present study therefore aims to establish the pharmacodynamic interaction between ezetimibe, Simvastatin and Omega 3 fatty acids and would offer significant clinical benefits, safe treatment for hyperlipidemia.

The primary objective of study was to evaluate the potential for pharmacodynamic interaction between Simvastatin, ezetimibe and omega 3 fatty acids in subjects with low-density lipoprotein cholesterol (LDL) ≥ 130 mg/dl and triglycerides ≥ 200 mg/dl and total cholesterol ≥ 200 mg/dl.

A secondary objective of this study was to evaluate the safety and tolerability of this treatment, with careful monitoring of enzymes assessing muscle and liver injury.

In detail, the aims of the study described here were as set out below:

4.1 STUDY DESIGN AND DATA HANDLING

It was proposed to conduct a placebo controlled, randomized, parallel group study in which subjects were randomized and grouped. Subjects received either selected drugs alone or in different combinations.

4.2 MEASUREMENT OF SAFETY AND VITAL SIGNS

Propose to collect blood samples on 0 day (before dosing), 25th day, 50th day and 90th day for monitoring signs of muscle and liver injury. Vital signs (body mass index, AST, ALT, CPK, blood pressure and heart rate were monitored during screening, before treatment administration and at 25th day, 50th day and 90th day. subjects were continually observed and questioned for possible adverse events.

4.3 PHARMACODYNAMICS

Propose to collect blood samples on 0th day (before dosing), 25th day, 50th day and 90th day for the measurement of,

1. Total Cholesteral (TC)
2. Triglycerides (TG)
3. Low Density Lipoproteins (LDL)
4. High Density Lipoproteins (HDL) Serum Glutamate Pyruvate Transaminase (SGPT)
5. Serum Glutamate Oxaloacetate Transaminase (SGOT)
6. Creatine Kinase (CK)

The concentrations were determined by direct quantitative assay methods (enzymatic colorimetric tests) using validated commercial assay kits.

4.4 STATISTICAL ANALYSIS

Propose to calculate mean, standard deviation or standard error, ANOVA and dunnett's test for the lipid parameters TC, TG, HDL and. LDL,

5. METHODOLOGY

5.1 STUDY DESIGN AND DATA HANDLING

The study was a comparative, prospective, randomized open label study. The study was conducted during the period 2011-2012.

The subjects were selected from the panel of subjects enrolled with the Centre of Clinical Research, Roxaane Research Private limited, Chennai, seven days prior to the commencement of the study, subjects were screened based on the inclusion and exclusion criteria of the study.

On the basis of this preliminary screening, 192 subjects were selected. No concomitant medication was allowed during the study phase. Subjects were also instructed to refrain from consuming alcohol, smoking or other stimulant drinks during this period.

Subjects were stabilized as outsubjects on an NCEP Step I diet, study treatments were administered orally with 200 ml of water.

The protocol of the study was submitted to the Institutional Human Ethical Committee and the approval for conducting the same was obtained. Prior to the commencement of the study, each subject was provided with an information sheet giving details of the investigational drugs, procedure and potential risk involved and a written consent was obtained. They were instructed that they were free to withdraw their consent and to discontinue their participation in the study at any time without prejudice or reason.

The parameters were measured before the drug intervention and after drug intervention. The average of change in the each measurement between the groups was compared.

The data collected was entered into a specially designed profoma (Case Recording Form) for the study, which was then subjected to statistical analysis for independent groups. While registering the subjects for our study, the following inclusion and exclusion criteria were applied.

a) Inclusion criteria:

- Age : Between 20 – 70 years
- Sex : Both males and females
- LDL Cholesterol : > 100 mg/dl
- HDL Cholesterol : < 40 mg/dl
- Triglycerides : > 150 mg/dl
- Total Cholesterol : > 200 mg/dl
- Serum creatinine : < 1.2 mg/dl
- No previous history of using any antilipedemic drugs
- Normal liver function tests values
- Normal CPK level.

b) Exclusion criteria:

- Subjects with Renal failure
- Subjects with Hepatic failure
- Pregnant & lactating women
- Past and present use of drugs having reported interaction with Simvastatin
- Serum aminotransferase more than three times normal.
- Serum CPK more than three times normal.
- Heavy smoking.
- Dropout subjects were excluded.

Investigations

After giving a written informed consent, all the subjects included in the study were subjected to the following investigations.

- Routine laboratory tests: Complete blood count, Hemoglobin percentage, fasting blood sugar, serum creatinine, blood urea nitrogen, electrolyte, liver function test and CPK.
- Lipid profile.
- ECG.

Study Visits

The subjects who had passed the screening laboratory tests and met inclusion criteria were inducted in the study. All the subjects were advised to take low fat diet and exercise regularly. They were randomly allocated into eight groups for participation in a randomized study. At each visit subjects were fully inquired about drug compliance, and side effects of the drugs. Subjects were also motivated to keep their nutritional habits, physical activity, and good life style as consistent as possible throughout the study period.

Subjects received one of the following eight treatments, daily for 90 days:

- Group I: Placebo
- Group II: Simvastatin 20 mg
- Group III: Ezetimibe 10 mg
- Group IV: Omega 3 fatty acids 4 g,
- Group V: Simvastatin 20 mg and Ezetimibe 10 mg
- Group VI: Simvastatin 20 mg and Omega 3 fatty acids 4 g
- Group VII: Ezetimibe 10 mg and Omega 3 fatty acids, 4 g
- Group VIII : Simvastatin 20 mg , Ezetimibe 10 mg and Omega 3 fatty acids 4 g

All the subjects were made to assemble in the Center of Clinical Research the subjects were given code numbers and allocated to the treatment in accordance with the randomized code. Their pulse rates and blood pressures were recorded and disposable needles were used with strict aseptic precautions for blood collection. Fasting Blood samples (10 ml) were collected using disposable syringes in pre-heparinised centrifugal tubes at 0th (before drug administration), 25th, 50th and 90th days. The samples were centrifuged at 3500 rpm for 10 minutes to separate plasma. They were transferred into airtight containers and stored at -20^{0} C until starting of analysis.

The study was monitored by a physician and a clinical pharmacologist. The subjects were monitored for abnormal symptoms during the study period and for 15 days after the study period and if noticed, the details were entered in the case report sheets and tabulated at the end of the study.

Lipid profile was done at 0th day, 25th day, 50th day and 90th day. Liver function test was done at 25th day, 50th day and at the end of the study. Discontinuation criteria included persistent 3-fold the upper limit of normal (ULN) increases in ALT or AST and 10-fold the ULN for CK with or without muscular symptoms or 5- to 10-fold increases in CK with symptoms.

5.2 MEASUREMENT OF SAFETY AND VITAL SIGNS

Blood samples were collected on 0th day, 25th day, 50th day and 90th day and signs of muscle and liver injury were monitered. Vital signs (body mass index, AST, ALT,CPK, blood pressure and heart rate were monitored during screening, before treatment administration and at 25th day, 50th day and 90th day after the treatment. Subjects were continually observed and questioned for possible adverse events.

5.2.1 Estimation of Serum Glutamate Pyruvate Transaminase (SGPT)

The normal range of values for SGPT is about 5 to 40 units per liter of serum. It is found to be distributed mainly in the liver and to a lesser extend in the kidney and muscles. ALT level elevated in liver damage and myocardial infarction.

Serum glutamate pyruvate transaminase, SGPT also called as Alanine amino transferase (ALT) was determined by using Reitman and Frankel method [55].

Reagents Used

Sl. No.	Reagent composition	Conc. in the final test mixed
1.	Buffered Alanine α- KG substrate pH: 7.4	-
2.	DNPH colour reagent	-
3.	Sodium Hydroxide	4 N
4.	Working pyruvate standard	2 mM

SGPT converts l-alanine and α-keto glutarate to Pyruvate and glutamate. Pyruvate with 2, 4 DNPH resulted in brownish red colour complex in an alkaline medium.The colour intensity was directly proportional to the SGPT concentration in the serum and was measured photometrically at 505nm under alkaline condition.

5.2.2 Estimation of Glutamate Oxaloacetate Transaminase (SGOT)

The normal range of values for AST (SGOT) is about 5 to 45 units per liter of serum. AST level elevated in myocardial infarction, muscular dystrophy, and liver necrosis.

Serum oxaloacetate transaminase, SGOT also called as Aspartate amino transferase. AST was determined by using Reitman and Frankel method

Reagents Used

Sl. No.	**Reagent composition**	**Conc. in the final test mixed**
1.	Buffered Aspartate α-KG substrate pH: 7.4	-
2.	DNPH colour reagent	-
3.	Sodium Hydroxide	4 N
4.	Working pyruvate standard	2 mM

Aspartate amino transferase or SGOT in serum converts α-aspartate to Oxaloacetate and glutamate. Oxaloacetate, with DNPH resulted in brown colour which was measured under alkaline condition.

5.2.3 Estimation of Creatine Kinase [56]

Reagents:

1. **CK** assay buffer
2. CK substrate
3. ATP (lyophilized)
4. CK enzyme mix (lyophilized)
5. CK developer (lyophilized)
6. NADH standard (lyophilized)
7. Positive control (lyophilized)

Procedure

In creatine kinase activity colorimetric assay kit, creatine kinase converts creatine into phosphocreatine and ADP.The generated phosphocreatine and ADP reacts with ck enzyme mix to form an intermediate which reduces a colorless probe to a colored product with strong absorbance at 450 nm.

Normal values are usually between 60 and 174 IU/L. Elevation of CK is an indication of damage to muscle. Increased CK level is associated with many diseases such as myocardial infarction, muscular dystrophy, pulmonary infarction and brain tumors. Lowered CK can be an indication of alcoholic liver disease and rheumatoid arthritis.

5.3 PHARMACODYNAMICS

Blood samples were collected for the measurement of Total Cholesteral (TC), Triglycerides (TG), Low Density Lipoproteins (LDL), High Density Lipoproteins (HDL), Serum Glutamate Pyruvate Transaminase (SGPT), Serum Glutamate Oxaloacetate Transaminase (SGOT) and Creatine Kinase on 0th day (before dosing, baseline value), 25th day, 50th day and 90th day by direct quantitative enzymatic colorimetric tests using validated commercial assay kits, using auto analyser.

5.3.1 Estimation of Total Cholesterol (TC)

Enzymatic methods[57] used were the assays of choice for the measurement of cholesterol. Span diagnostic kit was used for the estimation of total cholesterol, which followed cholesterol oxidase/ peroxidase method.

This reagent mixed with 10 microlitre aliquot of serum, incubated under controlled conditions for color development and absorbance measured in the visible portion of the spectrum generally at about 505 nm. The reagents typically use a bacterial cholesteryl ester hydrolase to cleave cholesteryl esters into cholesterol and fatty acid. The 3-OH group of cholesterol was then oxidised to a ketone and H_2O_2 in an oxygen requiring reaction catalysed by cholesterol oxidase. H_2O_2 with phenol and 4 amino antipyrine in a peroxidase catalysed reaction formed a colored dye.

Reagents Used

Sl. No.	Reagent composition	Conc. in the final test mixed
1.	Good's butter (pH 6.7)	50 mmol/l
2.	Phenol	5 mmol/l
3.	4 - Aminoantipyrine	0.3 mmol/l
4.	Cholesterol esterase	$\geq$200 U/l
5.	Cholesterol oxidase	$\geq$100 U/l
6.	Peroxidase	$\geq$3 KU/l

The concentration of standard cholesterol used was 200mg/dl. Fresh clear and unhaemolysed serum was used for the estimation. The reaction mixtures were mixed well and incubated for 10 min at 37 ^{0}C. The absorbance of reaction mixtures at 505nm against reagent blank was taken. The readings were measured by using an auto analyser.

5.3.2 Estimation of Triglycerides (TG)

The first step[57] was the lipase catalysed hydrolysis of triglycerides to glycerol and fatty acids. Glycerol was then phosphorylated in an ATP-requiring reaction catalysed by glycerokinase to glycerophosphate and adenosine di phosphate. Glycerophosphate was then oxidised to dihydroxyacetone and H_2O_2 in a glycerophosphate oxidase catalysed reaction. H_2O_2 was measured as shown above.

Reagents Used

Sl. No.	Reagent composition	Conc. in the final test mixed
1.	Pipes butter	50 mmol/l
2.	4-Chlorophenol	5 mmol/l
3.	Mg^{2+}	5 mmol/l
4.	ATP	1 mmol/l
5.	Lipase	$\geq$ 5000 U/l
6.	Peroxidase	$\geq$1000 U/l
7.	Glycerol Kinase	$\geq$400 U/l
8.	Glycerol - 3- phosphate oxidase	$\geq$4000 U/l
9.	4- Aminoantipyrine (4-AAP)	0.4 mmol/l

The concentration of standard triglyceride used was 200 mg/dl. Fresh clear and unhaemolysed serum was used for the estimation. The reaction mixtures were mixed well and incubated for 10 min at 37^0 C. The absorbance of sample and standard were measured against reagent blank at 505 nm. The readings were measured by using an autoanalyser.

5.3.3 Estimation of High Density Lipoprotein (HDL)

The concentration of HDL in plasma was assessed[57, 58] by determining the concentration of cholesterol associated with HDL. Polyanions like dextran sulphate when added to an aliquot of plasma react with positively charged groups on lipoproteins and formed a precipitate of the non-HDL lipoproteins within 10 minutes at room temperature. This precipitate was removed by centrifugation and HDL cholesterol was measured enzymatically in the supernatant on an auto analyser.

Span diagnostic kit was used for estimation of HDL cholesterol, which followed Cholesterol oxidase/peroxidase method. The concentration of standard HDL-cholesterol used was 50 mg/dl. The reaction mixtures were mixed well and incubated for 10 min at 37^0C. The absorbance of test and standards was measured against the reagent blank at 505 nm. The readings were measured by using an auto analyser

5.3.4 Estimation of Low Density Lipoprotein (LDL)

Using the Total Cholesterol, HDL and VLDL values, the LDL cholesterol level was measured by [59] using friedewald equation

5.4 STATISTICAL ANALYSIS

The scale measurement of LDL-C, TC, TG and HDL-C were described by their mean ± standard error values. For each variable, the comparison among independent groups for their mean values at for three stages (day 0, day 25, day 50, and day 90) was done by one way ANOVA followed by dunnets t test. Results were taken as significant for $p \leq 0.05$, else it was non-significant.

The difference in the mean values of the two groups was regarded statistically significant if the p-value was equal to or less than 0.05 and non-significant (ns) if the p-value was more than 0.05.

6. RESULTS AND DISCUSSION

This chapter describes the experimental results in the form of Tables and Figures along with a detailed discussion.

6.1 STUDY DESIGN AND DATA HANDLING

The protocol of the study was submitted to the Institutional Human Ethical Committee and the approval for conducting the same was obtained. Prior to the commencement of the study, each subject was provided with an information sheet giving details of the investigational drugs, procedure and potential risk involved and a written consent was obtained.

Among the 255 subjects screened, 43 subjects were primarily excluded from randomization for failure to meet eligibility criteria, 9 subjects excluded due to withdrawal of consent and 11 subjects excluded due to loss of follow-up during the placebo/diet run-in phase. 192 subjects were selected for the study, among these 171 subjects successfully completed the study.

Among the total 171 subjects, 101 subjects were males and 70 were females. 21 subjects were between 20-30 years, 25 subjects were between 31-40 years, 26 subjects were between 41-50 years, 44 subjects were between 51-60 years and 55 subjects were more than 60 years age group. The distribution of patient demographics and NCEP ATP III risk categories were comparable across treatment groups. Nearly half of the

Subjects (46%) had CHD or a CHD risk equivalent, whereas 37% had multiple (z2) risk factors conferring a 10-year risk for CHD, and 17% of subjects had b2 risk factors.

During an initial 15 day washout phase of the study, the participants received no antihyperlipidaemic medication and were stabilized on the NCEP Step I diet. After washout phase, subjects were randomized to one of the eight treatment groups.

- Group I: Placebo
- Group II: Simvastatin 20 mg
- Group III: Ezetimibe 10 mg
- Group IV: Omega-3 fatty acids 4 g,
- Group V: Simvastatin 20 mg and Ezetimibe 10 mg
- Group VI: Simvastatin 20 mg and Omega-3 fatty acids 4 g
- Group VII: Ezetimibe 10 mg and Omega-3 fatty acids 4 g
- Group VIII : Simvastatin 20 mg , Ezetimibe 10 mg and Omega-3 fatty acids 4 g

The treatment regimen consisted of distribution of subjects, demographics and baseline characteristics were comparable among the eight treatment groups. Study treatments were administered orally with 200 ml of water, once-daily dosing for 90 consecutive days.

The primary goal of the present study is to compare the efficacy, safety and tolerability of the selected drugs in mono therapy and in combined therapy.

6.2 MEASUREMENT OF SAFETY AND VITAL SIGNS

6.2.1 Demographic values of subjects on 0th day of treatment

Number of subjects in each group and their mean age is presented in Table 6.2.1. Sex and age distribution of subjects in each group is presented in table 6.2.2 and table 6.2.3, respectively.

21 subjects were selected in the Placebo group (Group I), among this 3 subjects were in the 20–30 years age group, 4 subjects were in the 31–40 years age groups, 5 subjects were in the 41–50 years age group and 4 subjects were in 51-60 years age group and only 5 subjects were in more than 60 years age group. 13 subjects were males and 8 subjects were females.

Out of 20 subjects in the Simvastatin group (Group II), 2 subjects were in the 20–30 years age group, 2 subjects were in the 31–40 years age group, 3 subjects were in the 41–50 years age group and 6 were in 51-60 years and only 7 subjects were in more than 60 years age group. 14 subjects were males and 6 subjects were females.

Out of 22 subjects in the Ezetimibe group (Group III), 3 subjects were in the 20–30 years age group, 3 subjects were in the 31–40 years age groups, 3 subjects were in the 41–50 years age group and 6 were in 51-60 years and only 7 subjects were in more than 60 years age group. 12 subjects were males and 10 subjects were females.

Out of 20 subjects in the Omega-3 fattys acid group (Group IV), 2 subjects were in the 20–30 years age group, 2 subjects were in the 31–40 years age groups, 3 subjects were in the 41–50 years age group and 5 were in 51-60 years and only 8 subjects were in more than 60 years age group. 15 subjects were males and 5 subjects were females.

Out of 20 subjects in the Simvastatin and ezetimibe group (Group V), 2 subjects were in the 20–30 years age group, 3 subjects were in the 31–40 years age groups, 2 were in the 41–50 years age group and 5 were in 51-60 years and only 8 subjects were in more than 60 years age group. 11 subjects were males and 9 subjects were females.

Out of 22 subjects in the Simvastatin and Omega-3 fatty acids group (Group VI), 3 subjects were in the 20–30 years age group, 4 subjects were in the 31–40 years age groups, 3 were in the 41–50 years age group and 6 were in 51-60 years age group and only 6 subjects were in more than 60 years age group. 10 subjects were males and 12 subjects were females.

Out of 23 subjects in the Ezetimibe and Omega -3 fatty acids group (Group VII), 3 subjects were in the 20–30 years age group, 3 subjects were in the 31–40 years age group, 4 were in the 41–50 years age group and 6 were in 51-60 years age group and only 7 subjects were more than 60 years age group. 13 subjects were males and 10 were females.

Out of 23 subjects in the Simvastatin, Ezetimibe and Omega-3 fatty acids group (Group VIII), 3 subjects were in the 20–30 years age group, 4 subjects were in the 31–40 years age group, 3 were in the 41–50 years age group and 6 were in 51-60 years age group and only 7 subjects were in more than 60 years age group. 13 subjects were males and 10 subjects were females.

6.2.2 Estimation of SGPT, SGOT and CK levels

The mean serum concentration of Serum Glutamate Pyruvate Transaminase (SGPT), Serum Glutamate Oxaloacetate Transaminase (SGOT) and creatine kinase (CK) at 0th day, 25th day, 50th day and 90th day for Group I to VIII is presented in Table 6.2.4.

No significant changes were observed for SGPT and SGOT levels when compared to baseline values with all the treatement groups at 25th day, 50th day and 90th day of treatment. No incidence of consecutive elevation of liver transaminase levels (SGPT and SGOT) more than three times the upper limit of the normal range (ULN) levels. No elevation of CK level of more than 10 times the ULN.

Summary of safety data is given in table 6.2.5. No serious adverse event was observed in any group of subjects. Each one subject from Simvastatin and Omega-3 fatty acids group showed allergic rash and three subjects from Simvastatin group and one subject from ezetimibe group were seen with gastrointestinal disturbances.

The mean values of Body Mass Index (BMI), heart rate, Systolic Blood Pressure (SBP) and Diastolic Blood Pressure (DBP) at 0th day, 25th day, 50th day and 90th day for Group I to VIII subjects are presented in Table 6.2.6.

When compared to baseline characteristics, there was no significant reduction was observed in mean body mass index, systolic blood pressure, diastolic blood pressure and heart rate for all the groups at 25th day and 50th day of treatment. Body mass index and heart rate were reduced in all the groups except placebo (Group-I). There was a mild reduction seen in mean DBP in Group IV (Omega-3 fatty acids), Group V (Simvastatin and Ezetimibe), Group VI (Simvastatin with Omega-3 fatty acids group), Group VII (Ezetimibe with Omega-3 fatty acids group) and Group VIII (Ezetimibe and Simvastatin with Omega-3 fatty acids group). There was a mild reduction seen in mean SBP in Group VIII (Simvastatin, ezetimibe and Omega-3 fatty acids group)

6.3 PHARMACODYNAMICS

Blood Lipids levels at the baseline, 25th day, 50th day and 90th day after treatment with placebo is presented in Table 6.3.1 and in Figure 6.3.1

6.3.1 Changes in Lipid Levels in Placebo Group

Mean values of Low Density Lipoproteins (LDL), Total Cholesterol (TC), Triglycerides (TG) and High Density Lipoproteins (HDL) in milligram per decilitre (mg/dl) before and after the treatment period is presented in Table 6.3.1 and Figure 6.3.1, and mean percent changes from baseline for LDL, TG, TC and HDL after the treatment period is presented in Figure 6.3.2.

Reduction in LDL values were 0.22% , 0.37% and 0.15% on 25th day, 50th day and 90th day, respectively, compared with the base values. Reduction in TC values were 0.21%, 0.17% and 0% on 25th day, 50th day and 90th day, respectively. Reduction in TG level was 0%, 0.1% and 0.13% on 25th day, 50th day and 90th day, respectively. The change in HDL level was 0.99%, 1.31% and 0.43% on 25th day, 50th day and 90th day, respectively.

None of the subjects treated with placebo achieved the NCEP ATP III LDL, TC, TG and HDL target lipid levels after the treatement period.

6.3.2 Changes in Lipid Levels in Simvastatin Treatment Group

Mean values of LDL, TG, TC and HDL in mg/dl before and after the treatment period is presented in Table 6.3.2 and Figure 6.3.3 and mean percent changes from baseline for LDL, TG, TC and HDL after the treatment period is presented in Table 6.3.2 and Figure 6.3.4.

Reduction in LDL values were 4.68% ($p \leq 0.05$), 12.28% ($p \leq 0.05$) and 17.3% ($p \leq 0.05$) on 25th day, 50th day and 90th day of the treatement, respectively, compared with the base values. Reduction in TC values were 3.17%, 7.37% ($p \leq 0.05$) and 14.24% ($p \leq 0.05$) on 25th day, 50th day and 90th day of the treatement, respectively. Reduction in TG level was 5.08% ($p \leq 0.05$), 10.95% ($p \leq 0.05$) and 19.95% ($p \leq 0.05$) on 25th day, 50th day and 90th day of the treatement, respectively. Elevation of HDL levels were 5.15% ($p \leq 0.05$), 10.45% ($p \leq 0.05$) and 19.58% ($p \leq 0.05$) on 25th day, 50th day and 90th day, respectively.

None of the subjects treated with Simvastatin achieved the NCEP ATP III LDL, TC, TG target levels after the treatement period, except HDL level.

6.3.3 Changes in Lipid Levels in Ezetimibe Treatement Group

LDL, TG, TC and HDL values in mg/dl before and after the treatment period is presented in Table 6.3.3 and Figure 6.3.5 and mean percent changes from baseline for LDL, TG, TC and HDL after the treatment is presented in Table 6.3.3 and Figure 6.3.6.

Reduction in LDL values were 4.79%, 11.64% ($p \leq 0.05$) and 19.19% ($p \leq 0.05$) on 25th day, 50th day and 90th day, respectively, compared with the base values. Reduction in TC values were 2.20%, 5.23% ($p \leq 0.05$) and 11.75% ($p \leq 0.05$) on 25th day, 50th day and 90th day, respectively. Reduction in TG levels were 3.40% ($p \leq 0.05$), 7.06% ($p \leq 0.05$) and 11.74% ($p \leq 0.05$) on 25thday, 50th day and 90th day, respectively. The increase in HDL levels were 4.83% ($p \leq 0.05$), 8.30% ($p \leq 0.05$) and 16.20% ($p \leq 0.05$) on 25th day, 50th day and 90th day, respectively.

None of the subjects treated with Ezetimibe achieved the NCEP ATP III LDL, TC, TG and HDL target lipid levels after the treatement period.

6.3.4 Changes in Lipid Levels in Omega-3 Fatty Acids Treatement Group

LDL, TG, TC and HDL values in mg/dl after the treatment period is presented in Table 6.3.4 and Figure 6.3.7, mean percent changes from baseline for LDL, TG, TC and HDL is presented in Table 6.3.4 and Figure 6.3.8.

Reduction in LDL values were 4.38%, 8.05% ($p \leq 0.05$) and 9.60% ($p \leq 0.05$) on 25th day, 50th day and 90th day, respectively, compared with the base values. Reduction in TC values were 2.84%, 6.24% ($p \leq 0.05$)

and 7.75% ($p \leq 0.05$) on 25th day, 50th day and 90th day, respectively. Reduction in TG level was 5.52%, 11.60% ($p \leq 0.05$) and 14.76% ($p \leq 0.05$) on 25th day, 50th day and 90th day, respectively. The increase in HDL levels were 4.39%, 8.79% ($p \leq 0.05$) and 19.35% ($p \leq 0.05$) on 25th day, 50th day and 90th day, respectively.

None of the subjects treated with Omega-3 fatty acids achieved the NCEP ATP III LDL, TC, TG levels during the treatement period, except HDL level.

6.3.5 Changes in Lipid Levels in Simvastatin and Ezetimibe Treatment Group

LDL, TG, TC and HDL values in mg/dl before and after the treatment period is presented in Table 6.3.5 and Figure 6.3.9 and mean percent changes from baseline for LDL, TG, TC and HDL is presented in Table 6.3.5 and Figure 6.3.10.

Reduction in LDL values were 10.93% ($p \leq 0.05$), 22.19% ($p \leq 0.05$) and 41.97% ($p \leq 0.05$) on 25th day, 50th day and 90th day, respectively. Reduction in TC values were 5.46% ($p \leq 0.05$), 12.13% ($p \leq 0.05$) and 18.19% ($p \leq 0.05$) on 25th day, 50th day and 90th day, respectively. Reduction in TG level was 10.49% ($p \leq 0.05$), 20.86% ($p \leq 0.05$) and 32.07% ($p \leq 0.05$) on 25th day, 50th day and 90th day, respectively. The increase in HDL levels were 6.39%($p \leq 0.05$), 13.69% ($p \leq 0.05$) and 26.48% ($p \leq 0.05$) on 25th day, 50th day and 90th day, respectively.

Subjects treated with Simvastatin and Ezetimibe achieved the NCEP ATP III target TC, and HDL levels during the treatement period. LDL level reached near optimum.

6.3.6 Changes in Lipid Levels in Simvastatin and Omega-3 fatty acids Treatment Group

Mean values LDL, TG, TC and HDL in mg/dl after the treatment period is presented in table 6.3.6 and Figure 6.3.11, mean percent changes from baseline for LDL, TG, TC and HDL are shown in Table 6.3.6 and Figure 6.3.12.

Reduction in LDL values were 7.42% ($p \leq 0.05$), 15.60% ($p \leq 0.05$) and 24.80% ($p \leq 0.05$) on 25th day, 50th day and 90th day, respectively.. Reduction in TC values were 2.98%, 6.60% ($p \leq 0.05$) and 12.70% ($p \leq 0.05$) on 25th day, 50th day and 90th day, respectively. Reduction in TG levels were 11.56% ($p \leq 0.05$), 24.07% ($p \leq 0.05$) and 41.08% ($p \leq 0.05$) on 25th day, 50th day and 90th day, respectively. The increase in HDL levels were 5.06% ($p \leq 0.05$), 12.11% ($p \leq 0.05$) and 23.93%($p \leq 0.05$) on 25th day, 50th day and 90th day, respectively.

Subjects treated with Simvastatin and Omega-3 fatty acids achieved the NCEP ATP III target TG and HDL levels after the treatment period.

6.3.7 Changes in Lipid Levels in Ezetimibe and Omega-3 fatty acids Treatment Group

Mean values LDL, TG, TC and HDL in mg/dl after the treatment period are presented in Table 6.3.7 and Figure 6.3.13, mean percent changes from baseline for LDL, TG, TC and HDL are shown in Table 6.3.7 and Figure 6.3.14.

Reduction in LDL values were 4.02% ($p \leq 0.05$), 7.60% ($p \leq 0.05$) and 10.48% ($p \leq 0.05$) on 25th day, 50th day and 90th day, respectively. Reduction in TC values were 3.30% ($p \leq 0.05$), 7.36% ($p \leq 0.05$) and 10.96% ($p \leq 0.05$) on 25th day, 50th day and 90th day, respectively. Reduction in TG levels were 4.64% ($p \leq 0.05$), 10.84% ($p \leq 0.05$) and 15.67% ($p \leq 0.05$) on 25th day, 50th day and 90th day, respectively. The increase in HDL levels were 3.13% ($p \leq 0.05$), 5.49% ($p \leq 0.05$) and 9.49% ($p \leq 0.05$) on 25th day, 50th day and 90th day, respectively.

None of the subjects treated with ezetimibe and Omega-3 fattyacids achieved the NCEP ATP III target LDL, TC, TG and HDL levels during the treatment period.

6.3.8 Changes in Lipid Levels in Simvastatin, Ezetimibe and Omega-3 Fattyacids Treatment Group

Mean values LDL, TG, TC and HDL in mg/dl after the treatment period Table 6.3.8 and Figure 6.3.15 and mean percent changes from baseline for LDL, TG, TC and HDL are shown in Table 6.3.8 and Figure 6.3.16.

Reduction in LDL values were 12.24% ($p \leq 0.05$), 25.67% ($p \leq 0.05$) and 49.73% ($p \leq 0.05$) on 25th day, 50th day and 90th day, respectively. Reduction in TC values were 7.86% ($p \leq 0.05$), 16.55% ($p \leq 0.05$) and 27.97% ($p \leq 0.05$) on 25th day, 50th day and 90th day, respectively. Reduction in TG level was 37.38% ($p \leq 0.05$), 41.59% ($p \leq 0.05$) and 56.51% ($p \leq 0.05$) on 25th day, 50th day and 90th day, respectively. The increase in HDL levels were 30.95% ($p \leq 0.05$), 61.12% ($p \leq 0.05$) and 89.09% ($p \leq 0.05$) on 25th day, 50th day and 90th day, respectively.

Subjects treated with Simvastatin, Ezetimibe and Omega-3 fatty acids achieved all the NCEP ATP III LDL, TC, TG and HDL target levels after the treatement period. subjects attained all four specified (by NCEP ATP III guidelines) treatment targets like LDL level less than 100 mg/dl (optimal) , total cholesterol (TC) level less than 200 mg/dl, triglycerides (TG) level less than 150 mg/dl, and HDL level between 40-60 mg/dl .

6.4 SERUM LOW DENSITY LIPOPROTEINS CHOLESTEROL LEVEL

Mean serum LDL- cholesterol was significantly decreased in all the groups at the completion of 90 days except placebo group. In placebo group , LDL-C was 133 mg/dl and 132.8 mg/dl on 0th and 90th of the treatement, respectively, which shows that there was no changes in LDL-C level was observed during the study period.

In group II, LDL-C was 181.5 mg/dl on 0th day of the treatment, which was reduced to 150.1 mg/dl on 90th day of the treatment. The percentage reduction was 17.3.

In group III, LDL-C was 200.1 mg/dl on 0th day, which was reduced to 161.7 mg/dl on 90th day of treatment. The percentage reduction was 19.19.

In group 1V, LDL-C was 180.1 on 0th day of the treatment, which was reduced to 162.8 on 90th day of the treatment. The percentage reduction was 9.60.

In group V, LDL-C was 185.6 on 0th day of the treatment, which was reduced to 107.7, on 90th day of the treatment. The percentage reduction was 41.97.

In group VI, LDL-C was 196.7 on 0th day of the treatment, which was reduced to 147.9 on 90th day of the treatment. The percentage reduction was 24.80.

In group VII, LDL-C was 184 on 0th day of the treatment, which was reduced to 164.7 on 90th day of the treatment. The percentage reduction was 10.48.

In group VIII, LDL-C was 192.8 on 0th day of the treatment, which was reduced to 96.91 on 90th day of the treatment. The percentage reduction was 49.73.

Subjects treated with Simvastatin and Ezetimibe treatment (Group V) achieved the NCEP target LDL near optimum level.

Subjects treated with Simvastatin, Ezetimibe and Omega-3 fatty acids treatment (Group VIII), achieved the NCEP target LDL optimum level.

6.5 SERUM TOTAL CHOLESTEROL LEVEL

Mean serum total cholesterol was significantly decreased in all the groups at the completion of 90 days except placebo group. In placebo group I, TC was 232.4 mg/dl and 232.4 mg/dl on 0th and 90th of the treatement, respectively, which shows that no changes in TC level was observed during the study period.

In group II, TC was 235.9 on 0th day, which was reduced to 202.3 on 90th day of the treatment. The percentage reduction was 14.24.

In group III, TC was 227.2 on 0th day, which was reduced to 215.3 on 90th day. The percentage reduction was 11.75.

In group 1V, TC was 232.1 on 0th day, which was reduced to 214.1 on 90th day. The percentage reduction was 7.75.

In group V, TC was 234.1 on 0th day, which was reduced to 191.5 on 90th day. The percentage reduction was 18.19.

In group VI, TC was 234.5 on 0^{tth} day, which was reduced to 204.7 on 90th day. The percentage reduction was 12.70.

In group VII, TC was 236.1 on 0th day, which was reduced to 210.2 on 90th day. The percentage reduction was 10.96.

In group VIII, TC was 230.2 on 0th day, which was reduced to 165.8 on 90th day. The percentage reduction was 27.97.

After 90 days, subjects with Simvastatin, Ezetimibe and Omega-3 fatty acids treatment (Group VIII), and subjects with Simvastatin and Ezetimibe treatment (Group V) achieved the NCEP target desirable total cholesterol (TC) level.

6.6 SERUM TRIGLYCERIDE LEVEL

Mean serum triglycerides were significantly decreased in all the groups at the completion of 90 days except placebo group. In placebo group I, TG was 287 mg/dl and 287.4 mg/dl on 0th and 90th of the treatement, respectively, which shows that no changes in TG level was observed during the study period.

In group II, TG was 230 on 0th day, which was reduced to 184.1 on 90th day. The percentage reduction was 19.95.

In group III, TG was 243.5 on 0th day, which was reduced to 214.9 on 90th day. The percentage reduction was 11.74.

In group 1V, TG was 231.7 on 0th day, which was reduced to 197.5 on 90th day. The percentage reduction was 14.76.

In group V, TG was 253.5 on 0th day, which was reduced to 172.2 on 90th day. The percentage reduction was 32.07.

In group VI, TG was 222.2 on 0th day, which was reduced to 130.9 on 90th day. The percentage reduction was 41.08.

In group VII, TG was 264.7 on 0th day, which was reduced to 223.2 on 90th day. The percentage reduction was 15.67.

In group VIII, TG was 280.3 on 0th day, which was reduced to 121.9 mg/dl on 90th day. The percentage reduction was 56.51.

After 90 days, subjects with Simvastatin, Ezetimibe and Omega-3 fatty acids treatment (Group VIII), subjects with Simvastatin and Omega-3 fatty acids treatment (Group VI) achieved the NCEP target TG normal level.

6.7 SERUM HIGH DENSITY LIPOPROTEINS LEVEL

Mean serum HDL was significantly decreased in all the groups at the completion of 90 days except placebo group. In placebo group, HDL was 33.19 mg/dl and 33 mg/dl on 0th and 90th of the treatment, respectively, which shows that no change in HDL level was observed during the study period.

- In group II, HDL was 33.95 on 0th day, which was elevated to 40.6 on 90th day. The percentage elevation was 19.58
- In group III, HDL was 33.95 on 0th day, which was elevated to 39.45 on 90th day. The percentage elevation was 16.20
- In group 1V, HDL was 34.1 on 0th day, which was elevated to 40.7 on 90th day. The percentage elevation was 19.35
- In group V, HDL was 33.6 on 0th day, which was elevated to 42.5 on 90th day. The percentage elevation was 26.48
- In group VI, HDL was 34.18 on 0th day, which was elevated to 42.36 on 90th day. The percentage elevation was 23.93
- In group VII, HDL was 34.74 on 0th day, which was elevated to 38.04 on 90th day. The percentage elevation was 9.49.
- In group VIII, HDL was 33.57 on 0th day, which was elevated to 63.48 on 90th day. The percentage elevation was 89.09.
- Except subjects with Ezetimibe monotherapy (Group III) and subjects with Ezetimibe and Omega-3 fatty acids combination therapy (Group VII), all the other group subjects achieved NCEP target HDL normal level.

6.8 COMPARISON OF PERCENTAGE OF CHANGE IN LIPID LEVEL AT THE END OF 25 DAYS OF TREATMENT

The percentage change from baseline in LDL-C after 25 days of Simvastatin and Ezetimibe monotherapies were 4.68 and 4.79. This change was 10.93% after combination treatment of 20 mg Simvastatin and 10 mg Ezetimibe. Combination treatment showed a greater reduction (double the response) in LDL-C compared with monotherapies in 25 days.

Percentage changes in LDL observed after 25 days of Omega 3 fatty acids monotherapy was 4.38%. However, the percentage reduction in LDL after treatement with 20mg Simvastatin with Omega 3 fatty acids 4 g combination and 10mg Ezetimibe with Omega 3 fatty acids were 7.42 and 4.02. Combination of Simvastatin, Ezetimibe and Omega-3 fatty acids treatment showed additive response (12.24%) of all the three drugs monotherapies in reducing LDL of subjects on 25th day.

The percentage change from baseline in TC after 25days of Simvastatin and Ezetimibe monotherapies were 3.17 and 2.20.This change was 5.46% after combination treatment of 20mg Simvastatin and 10mg Ezetimibe.Combination treatment showed a greater reduction (double the response) in TC compared with monotherapies in 25days. Change that was observed after 25 days of Omega 3 fatty acids monotherapy was 2.84%.

The percentage reduction in TC after treatement with 20 mg Simvastatin with Omega 3 fatty acids combination and 10 mg Ezetimibe with Omega 3 fatty acids were 2.98 and 3.30. Combination treatments did not show considerable improvement in reduction of TC levels. Combination of Simvastatin, Ezetimibe and Omega-3 fatty acids treatment showed additive response (7.86%) of all the three drugs in reducing TC of subjects on 25thday.

The percentage change from baseline in TG after 25 days of Simvastatin and Ezetimibe monotherapies were 5.08 and 3.40.This change was 10.49% after combination treatment of 20 mg Simvastatin and 10 mg Ezetimibe. Combination treatment showed a greater reduction (double the response) in TG compared with monotherapies in 25 days. Percentage change observed after 25 days of Omega 3 fatty acids monotherapy (5.52%) was more than the Simvastatin or Ezetimibe mono therapies. However, the percentage reduction in TG after treatement with 20mg Simvastatin with Omega 3 fatty acids combination and 10mg Ezetimibe with Omega 3 fatty acids were 11.56 and 4.64. Except the Ezetimibe and Omega-3 fatty acids combination, other Combinations showed improvement in reduction of TG levels. Combination of Simvastatin, Ezetimibe and Omega-3 fatty acids treatment showed synergistic response (37.38%) of all the three drugs response in reducing TG of subjects on 25th day.

The percentage elevation from baseline in HDL after 25 days of Simvastatin and Ezetimibe monotherapies were 5.15 and 4.83. This change was 6.39% after combination treatment of 20mg Simvastatin and 10 mg ezetimibe. Combination treatment showed a similar elevation in HDL compared with monotherapies in 25 days. Percentage HDL changes observed after 25 days of Omega 3 fatty acids monotherapy was 4.39. However, the percentage elevation in HDL after treatement with 20mg Simvastatin with Omega 3 fatty acids combination and 10mg Ezetimibe with Omega 3 fatty acids combinations were 5.06 and 3.13. Combination of Simvastatin, Ezetimibe and Omega-3 fatty acids treatment showed synergistic response (30.95%) of all the three drugs in elevating HDL ofsubjects on 25th day.

6.9 COMPARISON OF PERCENTAGE OF CHANGE IN LIPID LEVEL AT THE END OF 50 DAYS TREATMENT

The percentage change from baseline in LDL-C after 50 days of Simvastatin and ezetimibe monotherapies were 12.28 and 11.64. This change was 22.19 after combination treatment of 20mg Simvastatin and 10mg Ezetimibe. Combination treatment showed a greater reduction (double the response) in LDL-C compared with monotherapies in 50 days. Percentage change observed after 50 days of Omega 3 fatty acids monotherapy was 8.05. However, the percentage reduction in LDL after treatment with 20mg Simvastatin with Omega 3 fatty acids 4 g combination and 10mg Ezetimibe with 4 g Omega 3 fatty acids was 15.60 and 7.60. Combination of Simvastatin, Ezetimibe and Omega-3 fatty acids treatment showed additive response (25.67%) of all the three drugs in reducing LDL of subjects on 50th day.

The percentage change from baseline in TC after 50 days of Simvastatin and Ezetimibe monotherapies were 7.37 and 5.23. This change was 12.13% after combination treatment of 20 mg Simvastatin and 10mg ezetimibe. Combination treatment showed a greater reduction (double the response) in TC compared with monotherapies in 50 days. Percentage change observed after 50 days of Omega 3 fatty acids monotherapy was 6.24. However, the percentage reduction in TC after treatement with 20 mg Simvastatin with Omega 3 fatty acids 4 g combination and 10 mg Ezetimibe with Omega 3 fatty acids 4 g were 6.69 and 7.36. Combination treatments did not show considerable improvement in reduction of TC levels. However Combination of Simvastatin, Ezetimibe and Omega-3 fatty acids treatment showed improved response (16.55%) of all the three drugs in reducing TC of subjectson 50th day.

The percentage change from baseline in TG after 50 days of Simvastatin and Ezetimibe monotherapies were 10.95 and 7.06. This change was 20.86% after combination treatment of 20mg Simvastatin and 10 mg Ezetimibe. Combination treatment showed a greater reduction (additive response) in TG compared with monotherapies in 50 days. Percentage change observed after 50 days of Omega 3 fatty acids monotherapy was 11.60. However, the percentage reduction in TG after treatement with 20mg Simvastatin with 4 g Omega 3 fatty acids combination and 10mg Ezetimibe with 4 g Omega 3 fatty acids was 24.87 and 10.84. Combination of Simvastatin, Ezetimibe and Omega-3 fatty acids treatment showed synergistic response (41.59%) of all the three drugs in reducing TG of subjects on 50th day.

The percentage change from baseline in HDL after 50 days of Simvastatin and Ezetimibe monotherapies were 10.45 and 8.30. This change was 32% after combination treatment of 20 mg Simvastatin and 10mg Ezetimibe. This combination treatment showed synergistic response in HDL elevation compared with monotherapies in 50 days. The percentage change observed after 50 days of Omega 3 fatty acids monotherapy was 8.79. However, the percentage elevation in HDL after treatment with 20 mg Simvastatin with 4 g Omega 3 fatty acids combination and 10 mg Ezetimibe with 4 g Omega 3 fatty acids were 12.11 and 5.49. Combination of Simvastatin, Ezetimibe and Omega-3 fatty acids treatment showed synergistic response (61.12%) of all the three drugs in elevating HDL of subjects on 50th day.

Combination of Simvastatin, Ezetimibe and Omega-3 fatty acids treatment showed synergistic improvement in response (73.59%) than the all three drugs, in elevating HDL of subjects (50th day).

6.10 COMPARISON OF PERCENTAGE OF CHANGE IN LIPID LEVEL AT THE END OF 90 DAYS TREATMENT

The percentage change from baseline in LDL-C after 90 days of Simvastatin and Ezetimibe monotherapies were 17.30 and 19.19.This change was 41.97 after combination treatment of 20mg Simvastatin and 10mg Ezetimibe. Combination treatment showed a greater reduction (double the response) in LDL-C compared with monotherapies in 90 days. Percentage change observed after 90 days of Omega 3 fatty acids monotherapy was 9.60. However, the percentage reduction in LDL after treatement with 20mg Simvastatin with Omega 3 fatty acids combination and 10mg Ezetimibe with Omega 3 fatty acids were 24.80 and 10.48. Combination of Simvastatin, Ezetimibe and Omega-3 fatty acids treatment showed additive response (49.73%) of all the three drugs in reducing LDL of subjects on 90th day.

The percentage change from baseline in TC after 90days of Simvastatin and Ezetimibe monotherapies were 14.24 and 11.75. This change was 18.19% after combination treatment of 20mg Simvastatin and 10mg Ezetimibe.Combination treatment showed a greater reduction in TC compared with monotherapies in 90 days.

Percentage change observed after 25days of Omega 3 fatty acids monotherapy was 7.75. However, the percentage reduction in TC after treatment with 20mg Simvastatin with Omega 3 fatty acids combination and 10mg Ezetimibe with Omega 3 fatty acids was 12.70 and 10.96. Combination of Simvastatin, Ezetimibe and Omega-3 fatty acids treatment showed additive response (27.97%) of all the three drugs in reducing TC of subjects on 90th day.

The percentage change from baseline in TG after 90 days of Simvastatin and Ezetimibe monotherapies were 19.95 and 11.75. This change was 32.07% after combination treatment of 20mg Simvastatin and 10mg Ezetimibe. Combination treatment showed a greater reduction (double the response) in TG reduction when compared with monotherapies in 90 days. The percentage change after 90 days of Omega 3 fatty acids monotherapy observed was 14.76. However, the percentage reduction in TG after treatement with 20mg Simvastatin plus 4 g Omega 3 fatty acids combination and 10mg Ezetimibe with 4 g Omega 3 fatty acids was 41.08 and 15.67. Combination treatment with 20mg Simvastatin with Omega 3 fatty acids showed a considerable improvement in reduction of TG levels compared with monotherapies. Combination of Simvastatin, Ezetimibe and Omega-3 fatty acids treatment showed synergistic response (56.51%) of all the three drugs in reducing TG of subjects on 90th day.

The percentage change from baseline in HDL after 90 days of Simvastatin and Ezetimibe monotherapies were 19.58 and 16.20. This change was 26.48% after combination treatment of 20mg Simvastatin and 10mg Ezetimibe. Combination treatment showed a slightly more elevation of HDL compared with monotherapies in 90 days. Percentage changes observed after 90 days of Omega 3 fatty acids monotherapy was 19.35. However, the percentage elevation in HDL after treatment with Simvastatin with Omega 3 fatty acids combination and Ezetimibe with Omega 3 fatty acids was 26.48 and 23.93. Combination of Simvastatin, Ezetimibe and Omega-3 fatty acids treatment showed synergistic response (89.09%) of all the three drugs in elevating HDL of subjects on 90th day.

After 90 days of treatment period, Combination of Simvastatin, Ezetimibe and Omega-3 fatty acids treatment group was the leading lipid (LDL, TC, TG and HDL) lowering group (group VIII).

6.11 COMPARISION OF LIPID LEVELS OF SIMVASTATIN PLUS EZETIMIBE COMBINED THERAPY WITH EZETIMIBE OR SIMVASTATIN OR OMEGA-3 FATTY ACIDS MONOTHERAPY

On 25th day of the treatment, Simvastatin plus Ezetimibe showed significant reduction in LDL-C with mean percentage difference of 6.25% than Simvastatin monotherapy, 6.14% than ezetimibe monotherapy and 6.55% than omega-3 fatty acids monotherapy.

On 50th day of the treatment, Simvastatin plus Ezetimibe treatement group showed significant reduction in LDL-C with mean percentage difference of 9.91% than Simvastatin monotherapy and 10.55% than Ezetimibe monotherapy and 14.14% than omega-3 fatty acids monotherapy.

On 90th day of Simvastatin plus Ezetimibe treatment group showed significant reduction in LDL-C with mean percentage difference of 24.67% than Simvastatin monotherapy and 22.78% than ezetimibe monotherapy and 32.37% than omega-3 fatty acids monotherapy.

On 25th day of the treatment, Simvastatin plus Ezetimibe showed significant reduction in TC with mean percentage difference of 2.29% than Simvastatin monotherapy, 3.26% than ezetimibe monotherapy and 2.62% than omega-3 fatty acids monotherapy.

On 50th day of the treatment, Simvastatin plus Ezetimibe treatment group showed significant reduction in TC with mean percentage difference of 4.76% than Simvastatin monotherapy, 6.9% than Ezetimibe monotherapy and 5.89% than omega-3 fatty acids monotherapy.

On 90th day of the treatment, Simvastatin plus Ezetimibe treatement group showed significant reduction in TC with mean percentage difference of 3.95% than Simvastatin monotherapy 6.44% than Ezetimibe monotherapy and 10.44% than omega-3 fatty acids monotherapy.

On 25th day of the treatment, Simvastatin plus Ezetimibe showed significant reduction in TG with mean percentage difference of 5.41% than Simvastatin monotherapy, 7.09% than Ezetimibe monotherapy and 4.97% than omega-3 fatty acids monotherapy.

On 50th day of the treatment, Simvastatin plus Ezetimibe treatement group showed significant reduction in TG with mean percentage difference of 9.91% than Simvastatin monotherapy and 13.8% than Ezetimibe monotherapy and 9.26% than omega-3 fatty acids monotherapy.

On 90th day of the treatment, Simvastatin plus Ezetimibe treatment group showed significant reduction in TG with mean percentage difference of 12.12% than Simvastatin monotherapy and 20.33% than Ezetimibe monotherapy and 17.31% than omega- 3 fatty acids monotherapy.

On 25th day of treatment, Simvastatin plus Ezetimibe showed Significant elevation in HDL with mean percentage difference of 1.24% than Simvastatin monotherapy and 1.56% than Ezetimibe monotherapy and 2% than omega-3 fatty acids monotherapy

On 50th day of treatment, Simvastatin plus Ezetimibe showed Significant elevation in HDL with mean percentage difference of 3.24% than Simvastatin monotherapy, 5.39% than Ezetimibe monotherapy 4.9% than omega-3 fatty acids monotherapy.

On 90th day of treatment, Simvastatin plus Ezetimibe showed significant elevation in HDL with mean percentage 6.9% than

Simvastatin monotherapy and 10.28% than ezetimibe monotherapy and 7.13% than omega-3 fatty acids monotherapy.

6.12 COMPARISION OF LIPID LEVELS OF SIMVASTATIN PLUS OMEGA-3 FATTY ACIDS COMBINED THERAPY WITH EZETIMIBE OR SIMVASTATIN OR OMEGA-3 FATTY ACIDS MONOTHERAPY

On 25th day of the treatment, Simvastatin plus omega-3 fatty acids showed significant reduction in LDL-C with mean percentage difference of 2.74% than Simvastatin monotherapy, 2.63% than ezetimibe monotherapy and 3.04% than omega-3 fatty acids monotherapy.

On 50th day of the treatment, Simvastatin plus omega-3 fatty acids showed significant reduction in LDL-C with mean percentage difference of 3.32% than Simvastatin monotherapy, 3.96% than ezetimibe monotherapy and 7.55% than omega-3 fatty acids monotherapy.

On 90th day of the treatment, Simvastatin plus omega-3 fatty acids showed significant reduction in LDL-C with mean percentage difference of 7.5% than Simvastatin monotherapy, 5.61% than ezetimibe monotherapy and 15.2% than omega-3 fatty acids monotherapy.

On 25th day of the treatment, Simvastatin plus omega-3 fatty acids showed significant reduction in TC with mean percentage difference of 0.19% than Simvastatin monotherapy, 0.78% than ezetimibe monotherapy and 3.04% than omega-3 fatty acids monotherapy.

On 50th day of the treatment, Simvastatin plus omega-3 fatty acids showed significant reduction in TC with mean percentage difference of 0.77% than Simvastatin monotherapy, 1.37% than ezetimibe monotherapy and 0.36% than omega-3 fatty acids monotherapy.

On 90th day of the treatment, Simvastatin plus omega-3 fatty acids showed significant reduction in TC with mean percentage difference of 1.54% than Simvastatin monotherapy, 0.95% than ezetimibe monotherapy and 4.95% than omega-3 fatty acids monotherapy.

On 25th day of the treatment, Simvastatin plus omega-3 fatty acids showed significant reduction in TG with mean percentage difference of 6.48% than Simvastatin monotherapy, 8.16% than ezetimibe monotherapy and 6.04% than omega-3 fatty acids monotherapy.

On 50th day of the treatment, Simvastatin plus omega-3 fatty acids showed significant reduction in TG with mean percentage difference of 13.12% than Simvastatin monotherapy, 17.01% than ezetimibe monotherapy and 12.47% than omega-3 fatty acids monotherapy.

On 90th day of the treatment, Simvastatin plus omega-3 fatty acids showed significant reduction in TG with mean percentage difference of 21.13% than Simvastatin monotherapy, 29.34% than ezetimibe monotherapy and 26.32% than omega-3 fatty acids monotherapy.

On 25th day of the treatment, Simvastatin plus omega-3 fatty acids showed significant elevation in HDL with mean percentage difference of 0.09% than Simvastatin monotherapy, 0.23% than ezetimibe monotherapy and 0.67% than omega-3 fatty acids monotherapy.

On 50th day of the treatment, Simvastatin plus omega-3 fatty acids showed slight changes in HDL with mean percentage difference of 1.66% than Simvastatin monotherapy, 3.75% than ezetimibe monotherapy and 3.32% than omega-3 fatty acids monotherapy.

On 90th day of the treatment, Simvastatin plus omega-3 fatty acids showed changes in HDL with mean percentage difference of 4.35% than Simvastatin monotherapy, 7.73% than ezetimibe monotherapy and 4.58% than omega-3 fatty acids monotherapy.

6.13 COMPARISION OF LIPID LEVELS OF EZETIMIBE PLUS OMEGA-3 FATTY ACIDS COMBINED THERAPY WITH EZETIMIBE OR SIMVASTATIN OR OMEGA-3 FATTY ACIDS MONOTHERAPY

On 25th day of the treatment, Ezetimibe plus omega-3 fatty acids showed significant reduction in LDL-C with mean percentage difference of 0.66% than Simvastatin monotherapy, 0.77% than ezetimibe monotherapy and 0.36% than omega-3 fatty acids monotherapy.

On 50th day of the treatment, Ezetimibe plus omega-3 fatty acids showed significant reduction in LDL-C with mean percentage difference of 4.68% than Simvastatin monotherapy, 4.04% than ezetimibe monotherapy and 0.45% than omega-3 fatty acids monotherapy.

On 90th day of the treatment, Ezetimibe plus omega-3 fatty acids showed significant reduction in LDL-C with mean percentage difference of 6.82% than Simvastatin monotherapy, 8.71% than ezetimibe monotherapy and 0.88% than omega-3 fatty acids monotherapy.

On 25th day of the treatment, Ezetimibe plus omega-3 fatty acids showed significant reduction in TC with mean percentage difference of 0.13% than Simvastatin monotherapy, 1.1% than ezetimibe monotherapy and 0.46% than omega-3 fatty acids monotherapy.

On 50th day of the treatment, Ezetimibe plus omega-3 fatty acids showed significant reduction in TC with mean percentage difference of 0.01% than Simvastatin monotherapy, 2.13% than ezetimibe monotherapy and 1.12% than omega-3 fatty acids monotherapy.

On 90th day of the treatment, Ezetimibe plus omega-3 fatty acids showed significant reduction in TC with mean percentage difference of 3.28% than Simvastatin monotherapy, 0.79% than ezetimibe monotherapy and 3.21% than omega-3 fatty acids monotherapy.

On 25th day of the treatment, Ezetimibe plus omega-3 fatty acids showed significant reduction in TG with mean percentage difference of 0.44% lesser than Simvastatin monotherapy, 1.24% more than ezetimibe monotherapy and 0.88% more than omega-3 fatty acids monotherapy.

On 50th day of the treatment, Ezetimibe plus omega-3 fatty acids showed significant reduction in TG with mean percentage difference of 0.11% lesser than Simvastatin monotherapy, 3.78% more than ezetimibe monotherapy and 0.76% lesser than omega-3 fatty acids monotherapy. On 90th day of the treatment, Ezetimibe plus omega-3 fatty acids showed significant reduction in TG with mean percentage difference of 4.28% than Simvastatin monotherapy, 3.93% than ezetimibe monotherapy and 0.91% than omega-3 fatty acids monotherapy.

On 25th day of the treatment, Ezetimibe plus omega-3 fatty acids showed changes in HDL with mean percentage difference of 2.02% lesser than Simvastatin monotherapy, 1.7% lesser than ezetimibe monotherapy and 1.26% lesser than omega-3 fatty acids monotherapy.

On 50th day of the treatment, Ezetimibe plus omega-3 fatty acids showed changes in HDL with mean percentage difference of 4.96% lesser than Simvastatin monotherapy, 2.81% lesser than ezetimibe monotherapy and 3.3% lesser than omega-3 fatty acids monotherapy.

On 90th day of the treatment, Ezetimibe plus omega-3 fatty acids showed significant elevation in HDL with mean percentage difference of 10.09% lesser than Simvastatin monotherapy, 6.71% lesser than ezetimibe monotherapy and 9.86% than omega- 3 fatty acids monotherapy.

6.14 COMPARISION OF LIPID LEVELS OF SIMVASTATIN, EZETIMIBE PLUS OMEGA-3 FATTY ACIDS COMBINED THERAPY WITH EZETIMIBE OR SIMVASTATIN OR OMEGA-3 FATTY ACIDS MONOTHERAPY

On 25th day of the treatment, simvastatin, Ezetimibe plus omega-3 fatty acids showed significant reduction in LDL-C with mean percentage difference of 7.56% than Simvastatin monotherapy, 7.45% than ezetimibe monotherapy and 7.86% than omega-3 fatty acids monotherapy.

On 50th day of the treatment, simvastatin, Ezetimibe plus omega-3 fatty acids showed significant reduction in LDL-C with mean percentage difference of 13.39% than Simvastatin monotherapy, 14.03% than ezetimibe monotherapy and 17.62% than omega-3 fatty acids monotherapy.

On 90th day of the treatment, simvastatin, Ezetimibe plus omega-3 fatty acids showed significant reduction in LDL-C with mean percentage difference of 32.43% than Simvastatin monotherapy, 30.54% than ezetimibe monotherapy and 40.13% than omega-3 fatty acids monotherapy.

On 25th day of the treatment, simvastatin, Ezetimibe plus omega-3 fatty acids showed significant reduction in TC with mean percentage difference of 4.69% than Simvastatin monotherapy, 5.66% than ezetimibe monotherapy and 5.02% than omega-3 fatty acids monotherapy.

On 50th day of the treatment, simvastatin, Ezetimibe plus omega-3 fatty acids showed significant reduction in TC with mean percentage difference of 9.18% than Simvastatin monotherapy, 11.32% than ezetimibe monotherapy and 10.31% than omega-3 fatty acids monotherapy.

On 90th day of the treatment, simvastatin, Ezetimibe plus omega-3 fatty acids showed significant reduction in TC with mean percentage difference of 13.73% than Simvastatin monotherapy, 16.22% than ezetimibe monotherapy and 20.22% than omega-3 fatty acids monotherapy.

On 25th day of the treatment, simvastatin, Ezetimibe plus omega-3 fatty acids showed significant reduction in TG with mean percentage difference of 32.3% than Simvastatin monotherapy, 33.98% than ezetimibe monotherapy and 31.86% than omega-3 fatty acids monotherapy.

On 50th day of the treatment, simvastatin, Ezetimibe plus omega-3 fatty acids showed significant reduction in TG with mean percentage difference of 30.64% than Simvastatin monotherapy, 34.53% than ezetimibe monotherapy and 29.99% than omega-3 fatty acids monotherapy.

On 90th day of the treatment, simvastatin, Ezetimibe plus omega-3 fatty acids showed significant reduction in TG with mean percentage difference of 36.56% than Simvastatin monotherapy, 44.77% than ezetimibe monotherapy and 41.75% than omega-3 fatty acids monotherapy.

On 25th day of the treatment, simvastatin, Ezetimibe plus omega-3 fatty acids showed significant elevation in HDL with mean percentage difference of 25.8% than Simvastatin monotherapy, 26.12% than ezetimibe monotherapy and 26.56% than omega-3 fatty acids monotherapy.

On 50th day of the treatment, simvastatin, Ezetimibe plus omega-3 fatty acids showed significant elevation in HDL with mean percentage difference of 50.67% than Simvastatin monotherapy, 52.82% than ezetimibe monotherapy and 52.33% than omega-3 fatty acids monotherapy.

On 90th day of the treatment, simvastatin, Ezetimibe plus omega-3 fatty acids showed significant elevation in HDL with mean percentage difference of 69.51% than Simvastatin monotherapy, 72.89% than ezetimibe monotherapy and 69.74% than omega-3 fatty acids monotherapy.

Table 6.2.1 Numbers and Age of Subjects

Parameter	G-1	G-2	G-3	G-4	G-5	G-6	G-7	G-8
Number of subjects	21	20	22	20	20	22	23	23
Mean Age (Years)	46.38 ± 3.04	50.95 ± 2.90	49.14 ± 2.97	51.80 ± 2.92	56.20 ± 2.92	47.00 ± 2.88	48.04 ± 2.83	47.61 ±3.11

Table 6.2.2 Sex Distribution

SEX	G-1	G-II	G-III	G-IV	G-V	G-VI	G-VII	G-VIII	TOTAL
MALE	13 (61.90%)	14 (70%)	12 (54.54%)	15 (75%)	11 (55%)	10 (45.45%)	13 (56.52%)	13 (56.52%)	101 (59%)
FEMALE	8 (38%)	6 (30%)	10 (45.45%)	5 (25%)	9 (45%)	12 (54.54%)	10 (43.47%)	10 (43.47%)	70 (40.90%)
TOTAL	21 (100%)	20 (100%)	22 (100%)	20 (100%)	20 (100%)	22 (100%)	23 (100%)	23 (100%)	171 (100%)

Table 6.2.3 Age Distribution

AGE GROUP	G-1	G-II	G-III	G-IV	G-V	G-VI	G-VII	G-VIII	TOTAL
20-30	3	2	3	2	2	3	3	3	21
31-40	4	2	3	2	3	4	3	4	25
41-50	5	3	3	3	2	3	4	3	26
51-60	4	6	6	5	5	6	6	6	44
>60	5	7	7	8	8	6	7	7	55
TOTAL	21	20	22	20	20	22	23	23	171

Table 6.2.4 Mean SGPT, SGOT and CPK values

Parameter	Treatment			
	0th day	25th day	50th day	90th day
G-I				
SGPT (U/litre)	30.29 ± 0.72	25.95 ± 0.31	27.76 ± 0.56	35.00 ± 1.60
SGOT (U/litre)	25.81 ± 0.25	28.10 ± 0.56	26.86 ± 0.39	25.95 ± 0.31
CPK (IU/litre)	86.67 ± 1.95	76.33 ± 1.39	73.76 ± 0.93	73.76 ± 1.72
G-II				
SGPT (U/litre)	31.20 ± 0.54	31.45 ± 0.58	39.20 ± 1.28	41.80 ± 1.93
SGOT (U/litre)	26.45 ± 0.45	37.9 ± 1.65	37.75 ± 0.73	33.10 ± 1.19
CPK (IU/litre)	91.85 ± 2.38	96.90 ± 2.30	95.30 ± 1.65	92.35 ± 3.05
G-III				
SGPT (U/litre)	31.50 ± 0.72	32.55 ± 0.85	34.68 ± 0.82	39.86 ± 1.75
SGOT (U/litre)	26.82 ± 0.74	32.18 ± 0.80	33.55 ± 0.64	35.41 ± 1.14
CPK(IU/litre)	86.73 ± 1.17	100.2 ± 1.69	97.73 ± 2.24	83.05 ± 3.71
G-IV				
SGPT(U/litre)	31.50 ± 0.60	32.65 ± 0.96	37.45 ± 0.98	35.90 ± 1.62
SGOT(U/litre)	26.10 ± 0.31	39.25 ± 2.31	35.90 ± 1.28	35.80 ± 1.17
CPK(IU/litre)	88.15 ± 2.52	92.75 ± 1.63	91.30 ± 2.69	81.55 ± 2.98
G-V				
SGPT(U/litre)	31.45 ± 0.69	34.50 ± 1.16	35.85 ± 1.13	38.10 ± 1.20
SGOT(U/litre)	27.25 ± 0.58	38.35 ± 1.97	35.95 ± 0.68	35.41 ± 1.50
CPK(IU/litre)	86.95 ± 2.28	90.70 ± 1.37	97.00 ± 1.75	76.50 ± 2.90

Table 6.2.4 Mean SGPT, SGOT and CPK values (continued)				
G-VI				
SGPT(U/litre)	32.68 ± 0.73	34.91 ± 1.18	38.18 ± 1.08	38.09 ± 1.54
SGOT(U/litre)	26.68 ± 0.53	38.41 ± 1.92	34.64 ± 0.76	33.50 ± 1.50
CPK(IU/litre)	86.82 ± 2.17	88.36 ± 1.56	95.77 ± 2.02	81.41 ± 3.63
G-VII				
SGPT(U/litre)	32.52 ± 0.71	35.17 ± 1.02	39.91 ± 1.19	37.65 ± 1.35
SGOT(U/litre)	27.61 ± 0.54	34.35 ± 1.65	35.35 ± 1.11	35.39 ± 1.45
CPK(IU/litre)	88.09 ± 1.96	86.30 ± 1.33	97.43 ± 1.98	85.13 ± 4.01
G-VIII				
SGPT (U/litre)	32.74 ± 0.43	34.13 ± 0.94	35.96 ± 0.75	37.57 ± 1.79
SGOT (U/litre)	27.74 ± 0.71	35.26 ± 1.70	34.17 ± 0.90	34.78 ± 1.47
CPK (IU/litre)	88.17 ± 1.70	85.87 ± 1.81	83.22 ± 2.34	88.74 ± 4.15

SGPT - Serum glutamic pyruvic transaminase, **SGOT** - Serum glutamic oxaloacetic transaminase, **CPK** - Creatine Phosphokinase

Table 6.2.5 Summary of safety data

Adverse effect	G-I	G-II	G-III	G-IV	G-V	G-VI	G-VII	G-VIII
	N-	N-	N-	N-	N-	N-	N-	N-
Serious adverse effect	0	0	0	0	0	0	0	0
Death	0	0	0	0	0	0	0	0
Discontinued due to adverse effect	0	0	0	0	0	0	0	0
Allergic rash	0	1	0	1	0	0	0	0
Gastro intestinal related	0	3	1	0	0	0	0	0
Hepatitis related	0	0	0	0	0	0	0	0

N- = number of subjects

Table 6.2.6 Mean BMI, Heart rate, SBP and DBP values

Parameter	G-I			
	0th day	25th day	50th day	90th day
Body Mass Index	28.19 ± 0.50	27.05 ± 0.56	26.29 ± 0.29	26.76 ± 0.35
Heart rate	86.38 ± 2.22	70.81 ± 0.74	80.81 ± 1.03	70.90 ± 0.53
SBP	119.60 ± 1.14	122.2 ± 1.18	120.2 ± 0.61	121.8 ± 0.54
DBP	83.52 ± 1.08	82.90 ± 0.96	81.10 ± 0.43	80.0 ± 1.15
G-II				
Body Mass Index	29.40 ± 0.57	28.45 ± 0.74	25.80±0.32	25.05±0.19
Heart rate	88.85 ± 1.98	71.40 ± 0.66	83 ± 0.71	72.00 ± 0.77
SBP	121.5 ± 1.55	123.1 ± 1.04	122.1 ± 0.80	121.6 ± 0.71
DBP	81.90 ± 1.12	84.62 ± 0.84	81.95 ± 0.43	83.20 ± 0.73
G-III				
Body Mass Index	30.55 ± 0.65	26.95 ± 1.24	24.41±0.39	24.27±0.28
Heart rate	88.86 ± 1.85	72.05 ± 0.60	81.68 ± 1.00	72.45 ± 0.62
SBP	121.0 ± 1.42	124 ± 0.67	123.8 ± 1.14	119.2 ± 0.80
DBP	81.64 ± 0.94	84.55 ± 0.99	81.68 ± 0.48	81.50 ± 0.71
G-IV				
Body Mass Index	28.30 ± 0.51	26.80 ± 0.64	24.65 ± 0.37	24.25 ± 0.21
Heart rate	87.00 ± 2.44	72.25 ± 0.66	83 ± 1.51	71.00 ± 0.45
SBP	120.3 ± 1.35	119.1 ± 1.36	119.3 ± 0.60	119.9 ± 1.02
DBP	83.15 ± 0.96	81.85 ± 0.96	79.10 ± 0.52	76.55 ± 0.55
G-V				
Body Mass Index	29.85 ± 0.52	27.00 ± 0.98	25.60 ± 0.36	24.40 ± 0.31
Heart rate	86.55 ± 2.04	71.60 ± 0.75	81 ± 1.18	71.65 ± 0.46
SBP	121 ± 1.63	123.5 ± 1.07	122.0 ± 1.16	120.0 ± 0.85
DBP	81 ± 1.01	84.10 ± 0.98	81.60 ± 0.39	77.40 ± 0.57

Table 6.2.6 Mean BMI, Heart rate, SBP and DBP values (Continued)				
G-VI				
Body Mass Index	29.82 ± 0.52	27.36 ± 0.42	25.45 ± 0.35	24.27±0.27
Heart rate	87.05 ± 1.95	71.68 ± 0.67	82 ± 1.09	71.55 ± 0.40
SBP	119.2 ± 1.25	121.6 ± 0.98	122.4 ± 1.05	119.8 ± 0.88
DBP	81.86 ± 1.01	81.45 ± 0.76	81.55 ± 0.43	79.77 ± 1.21
G-VII				
Body Mass Index	30.04 ± 0.50	27.39 ± 0.64	25.35 ± 0.33	24.35 ± 0.24
Heart rate	91.35 ± 2.42	71.26 ± 0.65	80 ± 1.11	71.17 ± 0.46
SBP	119.10 ± 1.33	121.8 ± 1.02	122.5 ± 1.01	121.3 ± 0.71
DBP	81.87 ± 1.02	82.04 ± 0.75	81.83 ± 0.40	79.52 ± 0.85
G-VIII				
Body Mass Index	29.78 ±0.78	27.96 ± 1.46	25.39 ± 0.32	23.57 ± 0.13
Heart rate	89.30 ± 2.00	70.65 ± 0.72	80 ± 1.12	71.65 ± 0.31
SBP	120.00 ± 1.49	118.2 ± 1.02	118.2 ± 1.45	118.7 ± 0.70
DBP	81.43 ± 0.97	81.65 ± 0.89	78.65 ± 0.51	77.43 ± 0.59

SBP(Systolic Blood Pressure), DBP(Diastolic Blood Pressure)

Table 6.3.1 Efficacy and percentage changes of placebo on lipid profiles

Lipoproteins mg/dl	Placebo control			
	base	25days	50days	90days
LDL	133.0 ± 0.20	132.7 ± 0.17 (-0.22%)	132.5 ± 0.13 (-0.37%)	132.8 ± 0.20 (-0.15%)
TC	232.4 ± 0.11	232.9 ± 0.30 (0.21%)	232.8 ± 0.25 (0.17%)	232.4 ± 0.26 (0%)
TG	287.0 ± 0.17	287.0 ± 0.06 (0%)	286.7 ± 0.22 (-0.1%)	287.4 ± 0.18 (0.13%)
HDL	33.19 ± 0.16	32.76 ± 0.21 (-0.99%)	32.62 ± 0.10 (-1.31%)	33.00 ± 0.19 (-0.43%)

Values (mg/dl) were expressed as mean ± SEM of 21 subjects.

* Significant ($p \leq 0.05$),

LDL- Low Density Lipoprotein, TC- Total Cholesterol, TG- Triglycerides, HDL- High Density Lipoprotein.

mg/dl – milligram per decilitre.

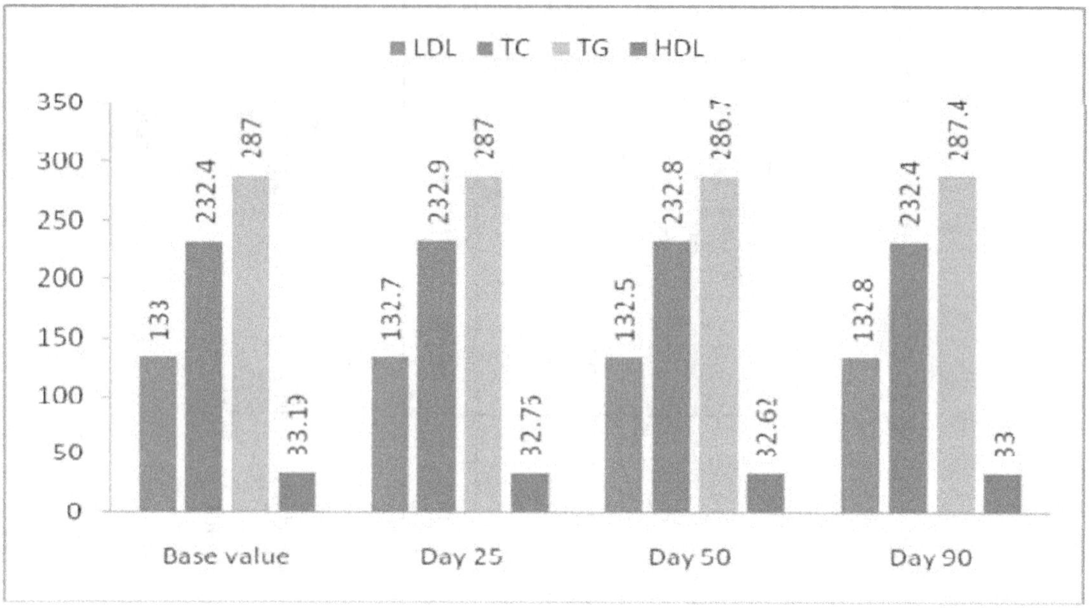

Figure 6.3.1 Efficacy of placebo on Lipid Profiles

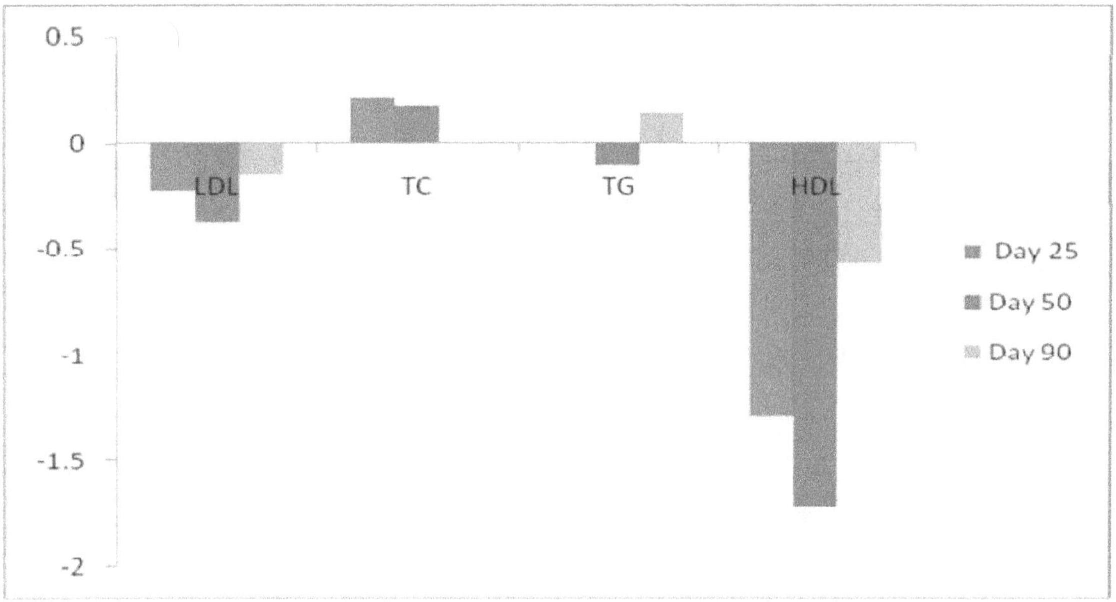

Figure 6.3.2 Percentage Change in Lipid Levels in Placebo Group

Table 6.3.2 Efficacy and percentage changes of Simvastatin on lipid profiles

Lipoproteins (mg/dl)	Base value	Day 25	Day 50	Day 90
LDL	181.50 ± 1.99	173.00 ± 1.97 (-4.68%)*	159.20 ± 1.95 (-12.28%)*	150.10 ± 2.18 (-17.3%)*
TC	235.90 ± 2.23	228.40 ± 2.06 (-3.17%)	218.50 ± 2.06 (-7.37%)*	202.30 ± 2.62 (-14.24%)*
TG	230.00 ± 2.78	218.30 ± 3.20 (-5.08%)*	204.80 ± 2.77 (-10.95%)*	184.10 ± 2.36 (-19.95%)*
HDL	33.95 ± 0.45	35.70 ± 0.34 (5.15%)*	37.50 ± 0.31 (10.45%)*	40.60 ± 0.44 (19.58%)*

Values (mg/dl) were expressed as mean ± SEM of 20 subjects.

* Significant ($p \leq 0.05$)

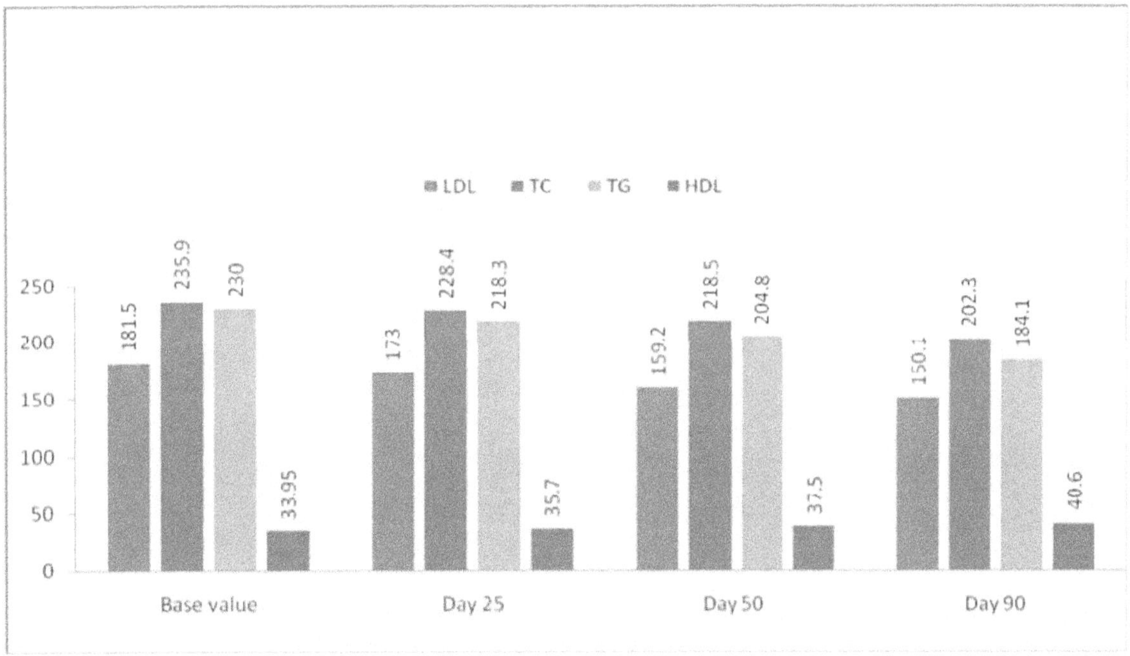

Figure 6.3.3 Efficacy of Simvastatin on lipid profiles

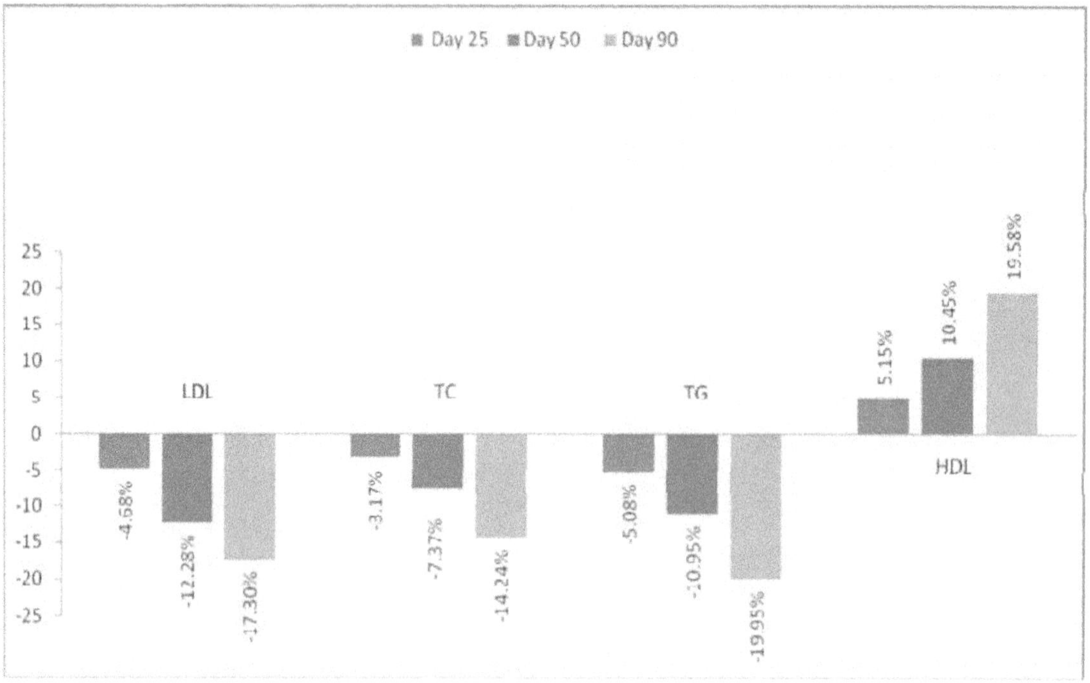

Figure 6.3.4 Percentage Change in Lipid Levels in Simvastatin Group

Table 6.3.3 Efficacy and Percentage Changes of Ezetimibe on lipid profiles

Lipoproteins (mg/dl)	Base value	Day 25	Day 50	Day 90
LDL	200.1 ± 2.69	190.5 ± 2.50 (-4.79%)	176.8 ± 2.86 (-11.64%)*	161.7 ± 3.31 (-19.19%)*
TC	227.2 ± 1.65	222.2 ± 1.48 (-2.20%)	215.3 ± 1.63 (-5.23%)*	200.5 ± 1.66 (-11.75%)*
TG	243.5 ± 1.75	235.2 ± 1.71 (-3.40%)	226.3 ± 2.08 (-7.06%)*	214.9 ± 1.82 (-11.74%)*
HDL	33.95 ± 0.39	35.59 ± 0.35 (4.83%)	36.77 ± 0.34 (8.30%)*	39.45 ± 0.39 (16.20%)*

Values (mg/dl) were expressed as mean ± SEM of 22 subjects

* Significant ($p \leq 0.05$)

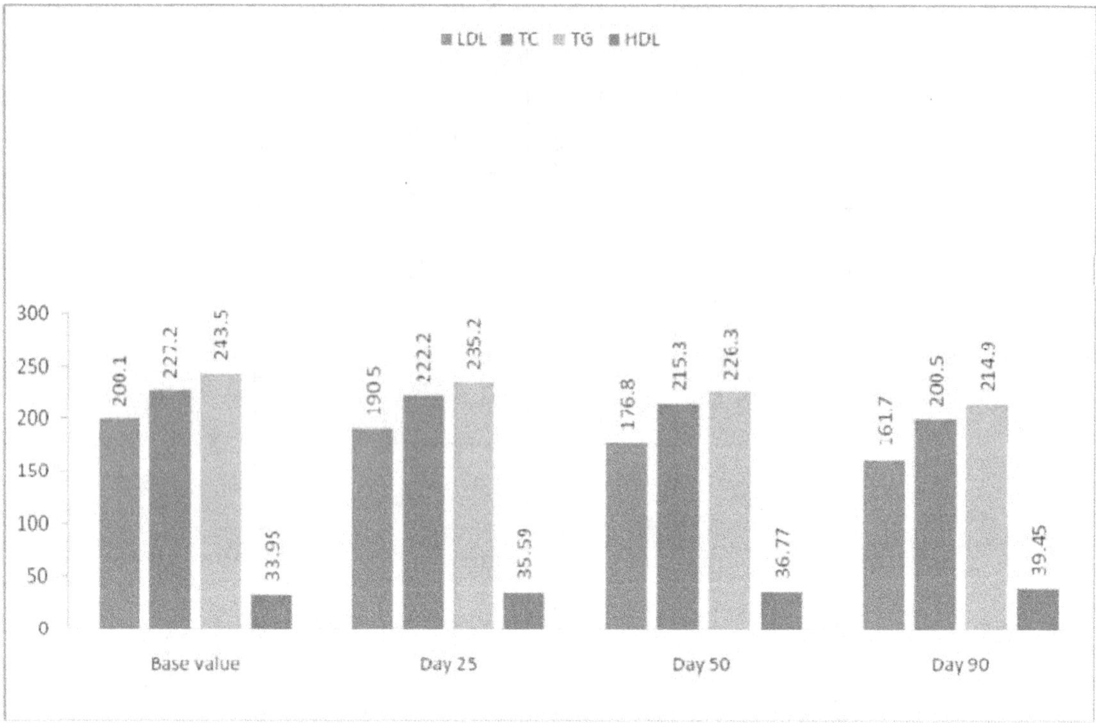

Figure 6.3.5 Efficacy of Ezetimibe on lipid profiles

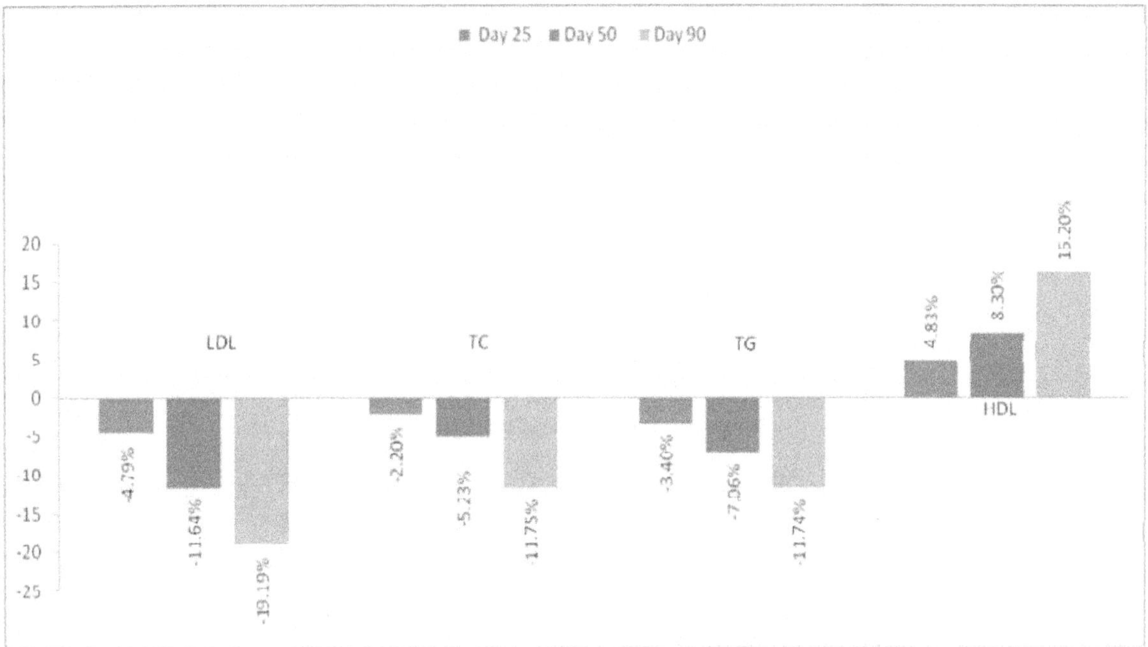

Figure 6.3.6 percentage Changes in Lipid Levels in Ezetimibe Group

Table 6.3.4 Efficacy and Percentage changes of Omega 3 fatty acids on lipid profiles

Lipoproteins (mg/dl)	Base value	Day 25	Day 50	Day 90
LDL	180.10 ± 2.80	172.20 ± 3.18 (-4.38%)	165.60 ± 3.15 (-8.05%)*	162.80 ± 2.78 (-9.60%)*
TC	232.10 ± 2.04	225.50 ± 2.09 (-2.84%)	217.60 ± 1.98 (-6.24%)*	214.10 ± 2.16 (-7.75%)*
TG	231.70 ± 4.41	218.90 ± 4.07 (-5.52%)	204.80 ± 4.51 (-11.60%)*	197.50 ± 5.11 (-14.76%)*
HDL	34.10 ± 0.57	35.60 ± 0.44 (4.39%)	37.10 ± 0.43 (8.79%)*	40.70 ± 0.51 (19.35%)*

Values (mg/dl) were expressed as mean ± SEM of 20 subjects.

* Significant ($p \leq 0.05$)

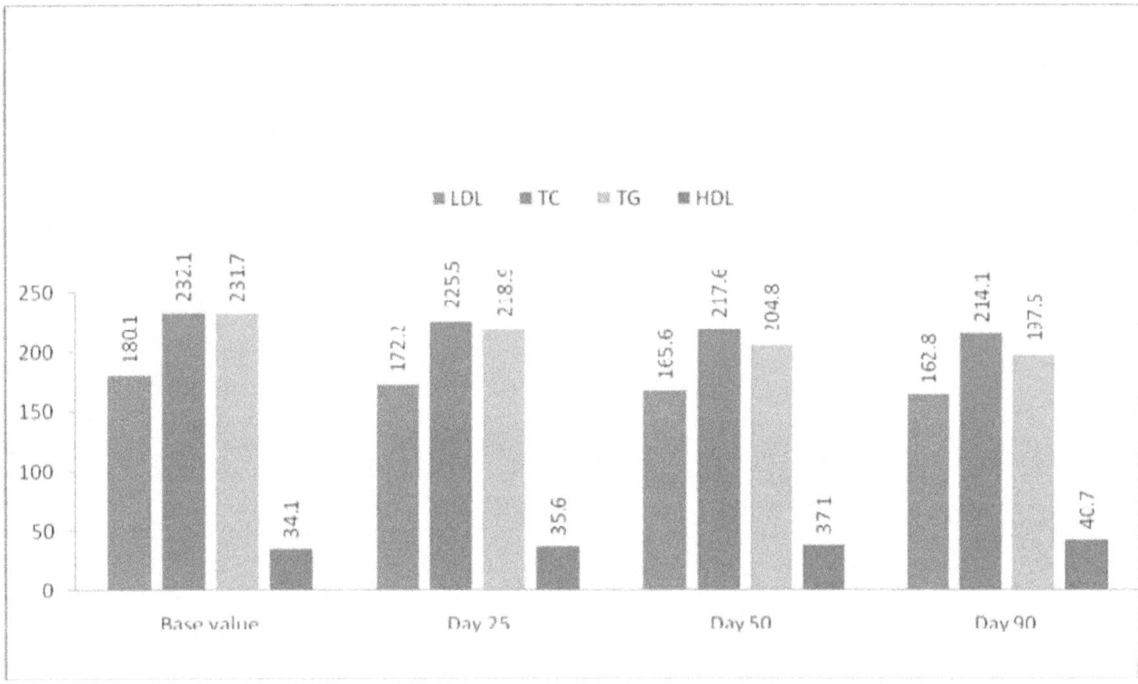

Figure 6.3.7 Efficacy of Omega- 3 fatty acids on lipid profiles

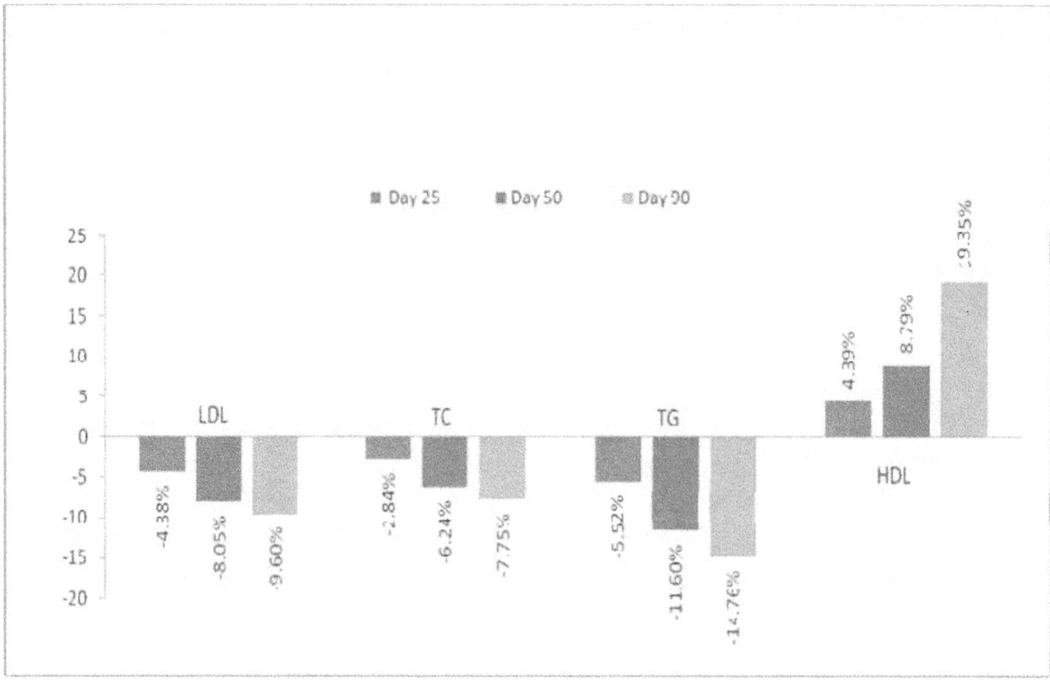

Figure 6.3.8 Percentage changes in Lipid Profiles in Omega-3 Fatty Acids Group

Table 6.3.5 Efficacy and percentage changes of Simvastatin and Ezetimibe on Lipid Profiles

Lipoproteins (mg/dl)	Base value	Day 25	Day 50	Day 90
LDL	185.60 ± 1.52	165.30 ± 1.79 (-10.93%)*	144.40 ± 1.73 (-22.19%)*	107.70 ± 1.79 (-41.97%)*
TC	234.10 ± 1.95	221.30 ± 2.16 (-5.46%)*	205.70 ± 2.96 (-12.13%)*	191.50 ± 2.52 (-18.19)*
TG	253.50 ± 3.39	226.90 ± 4.10 (-10.49)*	200.60 ± 4.16 (-20.86)*	172.20 ± 1.98 (-32.07%)*
HDL	33.60 ± 0.52	35.75 ± 0.48 (6.39%)*	38.20 ± 0.45 (13.69%)*	42.50 ± 0.45 (26.48%)*

Values (mg/dl) were expressed as mean ± SEM of 20 subjects.

*Significant ($p \leq 0.05$)

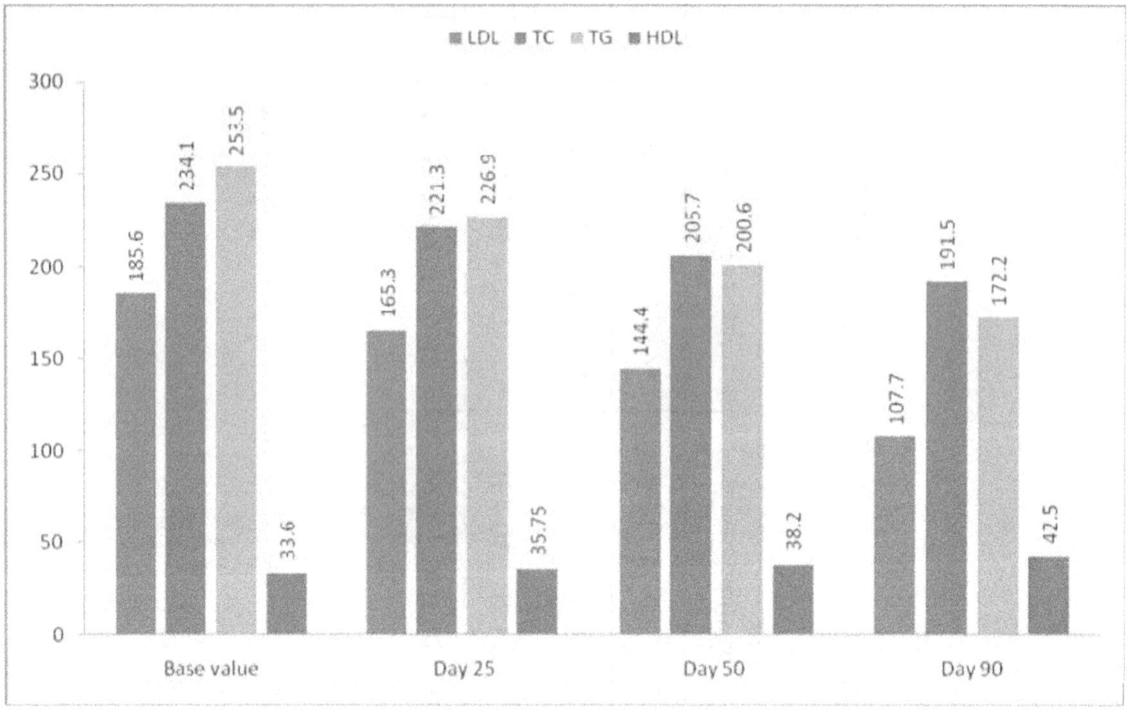

Figure 6.3.9 Efficacy of Simvastatin and Ezetimibe on Lipid Profiles

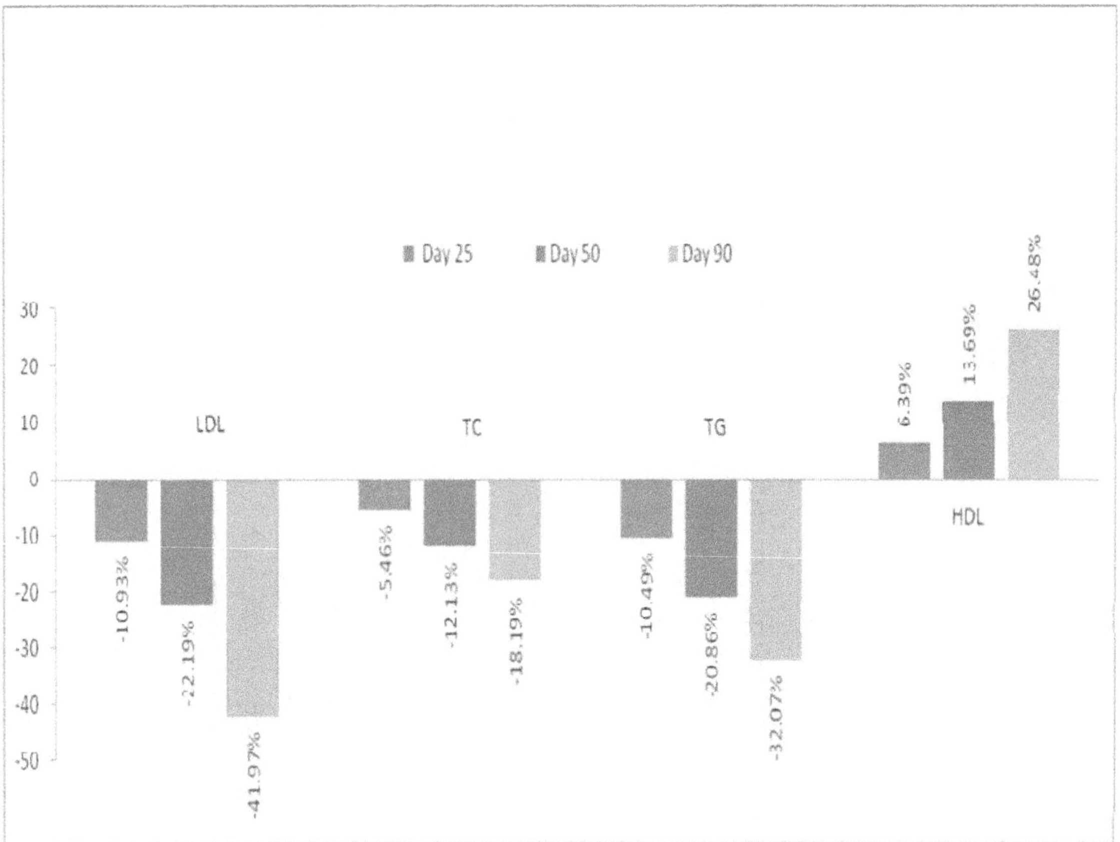

Figure 6.3.10 Percentage Changes in Lipid Levels in Simvastatin and Ezetimibe Group

Table 6.3.6 Efficacy and percentage changes of Simvastatin and Omega 3 fatty acids on lipid profiles

Lipoproteins (mg/dl)	Base value	Day 25	Day 50	Day 90
LDL	196.70 ± 2.56	182.10 ± 2.13 (-7.42%)*	166.00 ± 2.25 (-15.60%)*	147.90 ± 2.60 (-24.80%)*
TC	234.50 ± 2.11	227.50 ± 2.24 (-2.98%)	219.00 ± 2.19 (-6.60%)*	204.70 ± 1.86 (-12.70%)*
TG	222.20 ± 4.55	196.50 ± 3.99 (-11.56%)*	168.70 ± 4.36 (-24.07%)*	130.90 ± 4.08 (-41.08%)*
HDL	34.18 ± 0.45	35.91 ± 0.43 (5.06%)*	38.32 ± 0.34 (12.11%)*	42.36 ± 0.35 (23.93%)*

Values (mg/dl) were expressed as mean ± SEM of 22 subjects.

* Significant ($p \leq 0.05$)

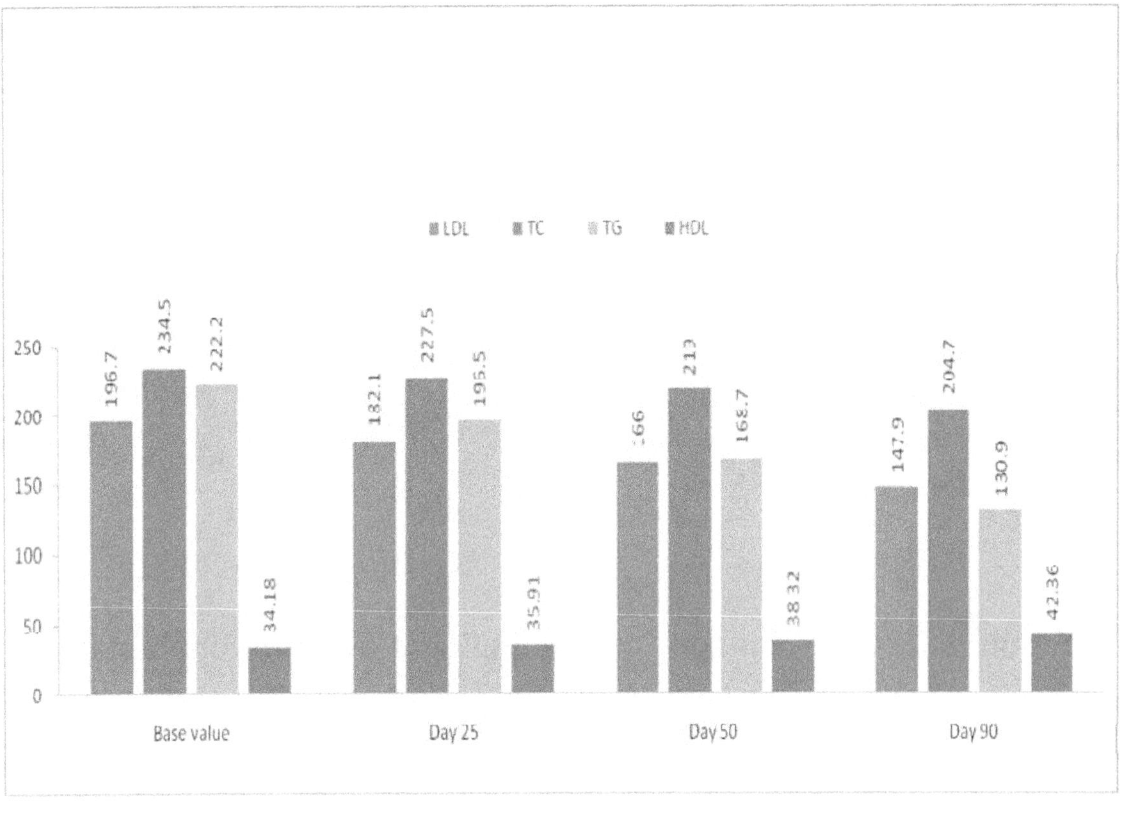

Figure 6.3.11 Efficacy of Simvastatin and Omega- 3 fatty acids on lipid profiles

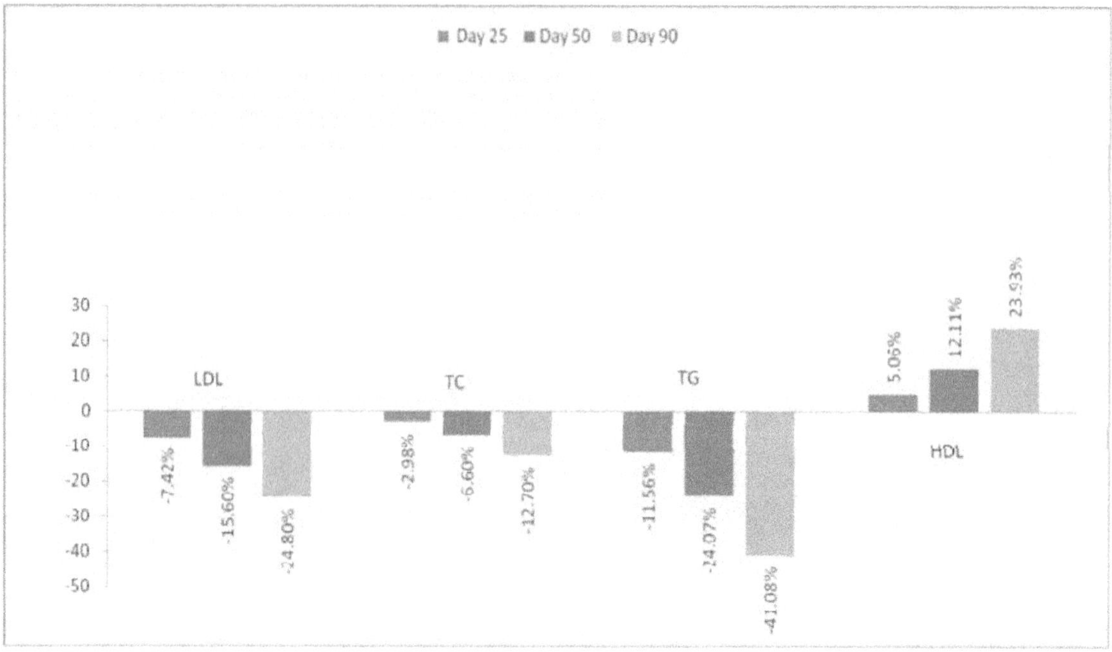

Figure 6.3.12 Percentage Change in lipid Levels in Simvastatin and Omega- 3 Fatty Acids Group

Table 6.3.7 Efficacy and percentage changes of Ezetimibe and Omega 3 Fatty Acids on Lipid Profiles

Lipoproteins (mg/dl)	Base value	Day 25	Day 50	Day 90
LDL	184.00 ± 2.20	176.60 ± 2.06 (-4.02%)*	170.00 ± 1.64 (-7.60%)*	164.70 ± 2.26 (-10.48%)*
TC	236.10 ± 1.51	228.30 ± 1.70 (-3.30%)*	218.70 ± 1.73 (-7.36%)*	210.20 ± 1.63 (-10.96%)*
TG	264.70 ± 1.83	252.40 ± 1.75 (-4.64%)*	236.00 ± 1.37 (-10.84%)*	223.20 ± 1.82 (-15.67%)*
HDL	34.74 ± 0.32	35.83 ± 0.29 (3.13%)*	36.65 ± 0.30 (5.49%)*	38.04 ± 0.29 (9.49%)*

. Values (mg/dl) were expressed as mean±SEM of 23 subjects.

* Significant ($p \leq 0.05$)

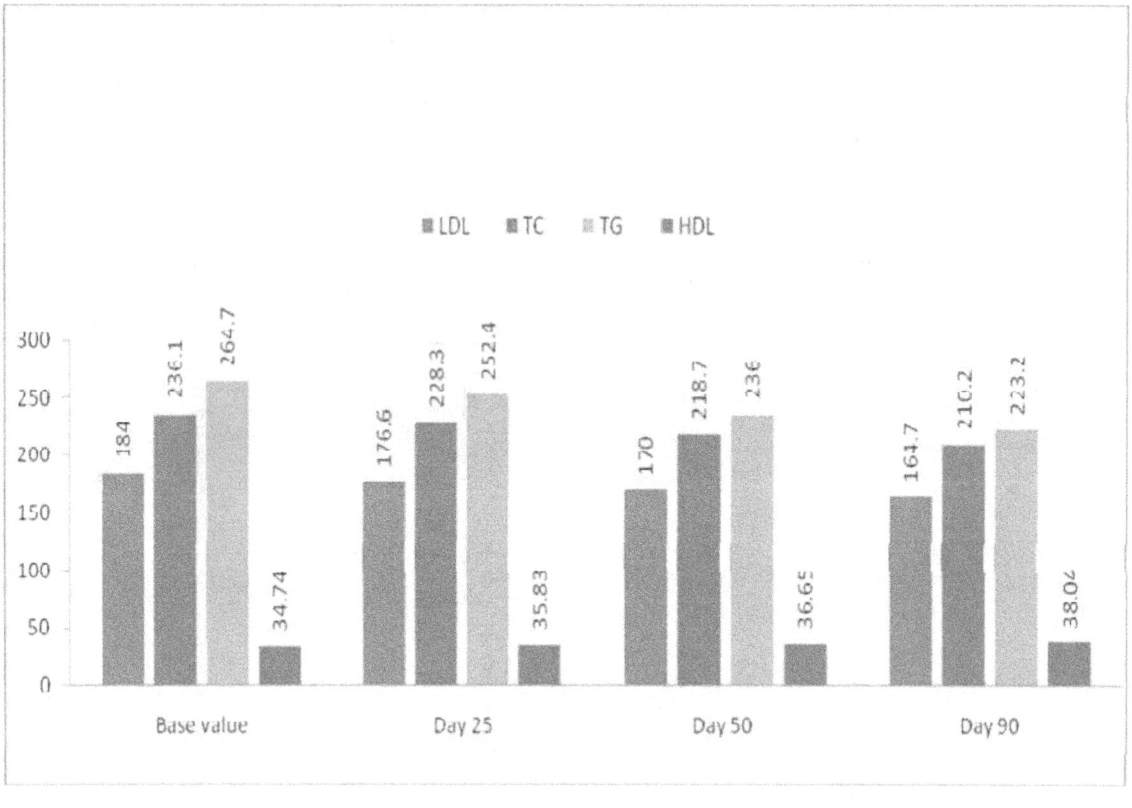

Figure 6.3.13 Efficacy of Ezetimibe and Omega-3 Fatty acids on Lipid Profiles

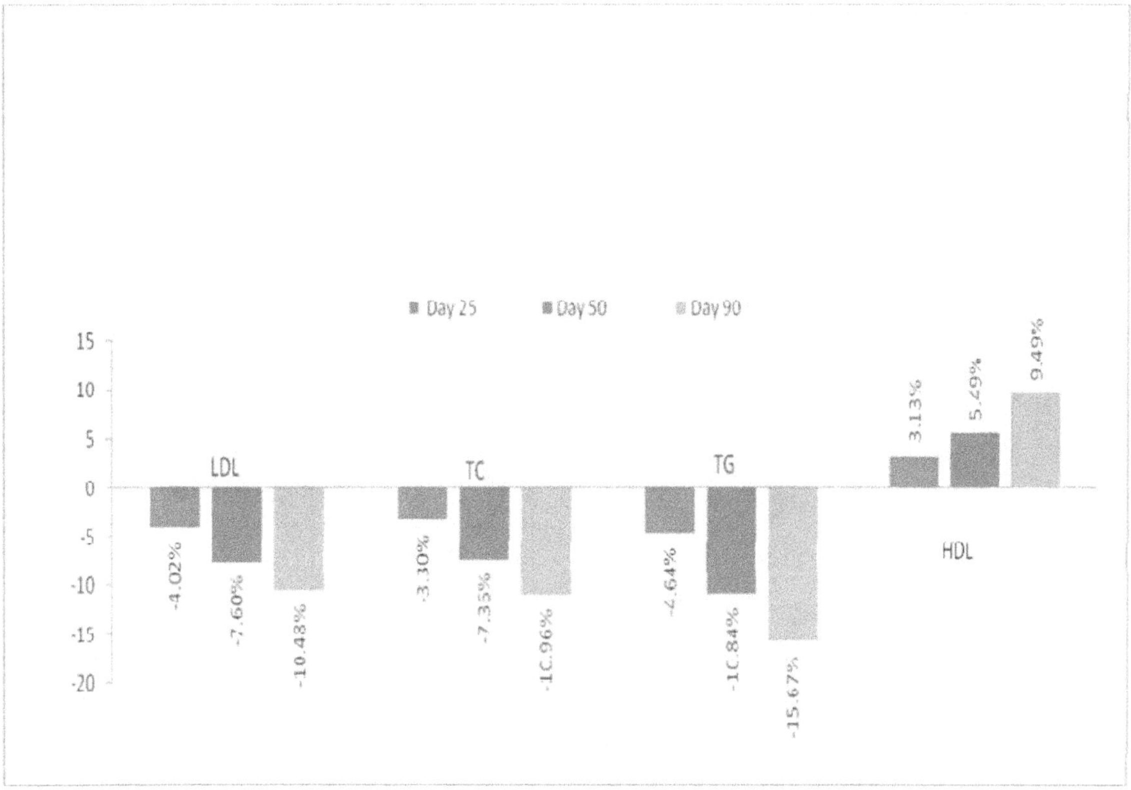

Figure 6.3.14 percentage Changes in lipid levels in Ezetimibe and Omega- 3 Fatty acids Group

Table 6.3.8 Efficacy and percentage changes of Simvastatin, Ezetimibe and Omega 3 Fatty Acids on Lipid Profiles

Lipoproteins (mg/dl)	Base value	Day 25	Day 50	Day 90
LDL	192.80 ± 3.48	169.20 ± 3.66 (-12.24%)*	143.30 ± 2.67 (-25.67%)*	96.91 ± 2.78 (-49.73%)*
TC	230.20 ± 2.15	212.10 ± 2.13 (-7.86%)*	192.10 ± 2.71 (-16.55%)*	165.80 ± 2.48 (-27.97%)*
TG	280.30 ± 1.21	175.50 ± 1.25 (-37.38%)*	163.70 ± 1.61 (-41.59%)*	121.90 ± 1.30 (-56.51%)*
HDL	33.57 ± 0.54	43.96 ± 0.57 (30.95%)*	54.09 ± 0.72 (61.12%)*	63.48 ± 0.95 (89.09%)*

Values (mg/dl) were expressed as mean ± SEM of 23 subjects

*Significant ($p \leq 0.05$)

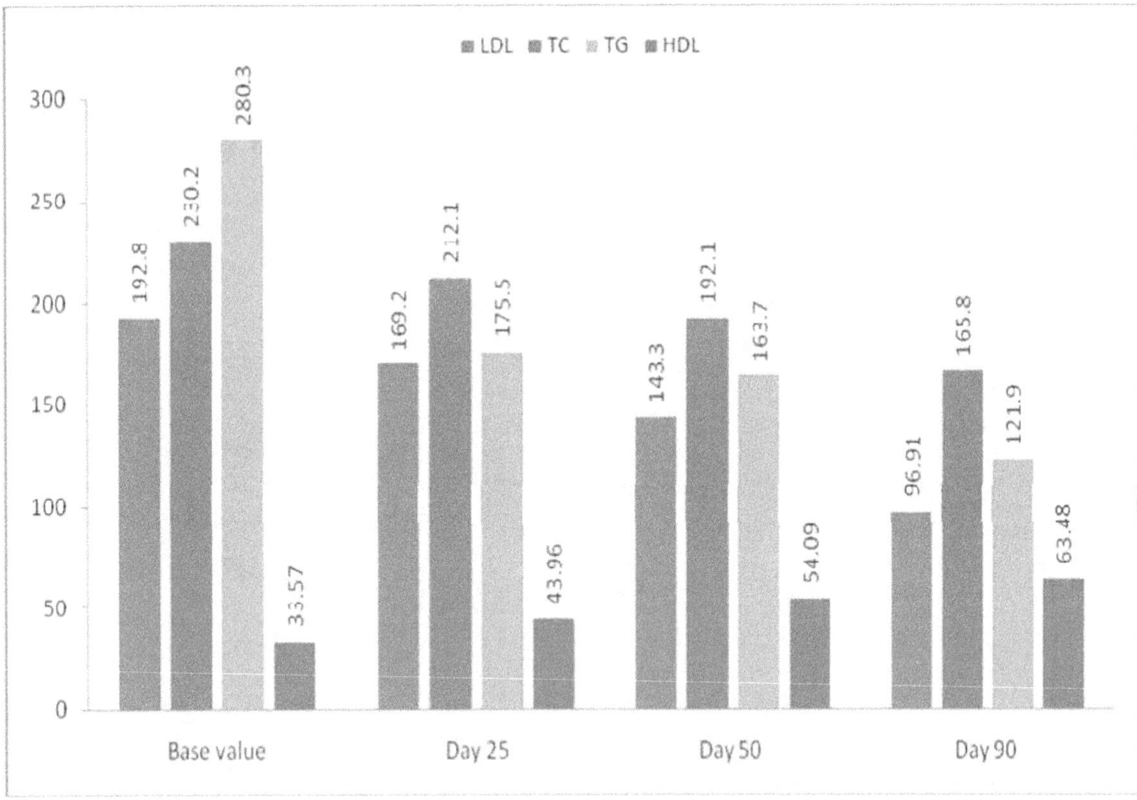

Figure 6.3.15 Efficacy of Simvastatin, Ezetimibe and Omega- 3 fatty Acids on Lipid Profiles

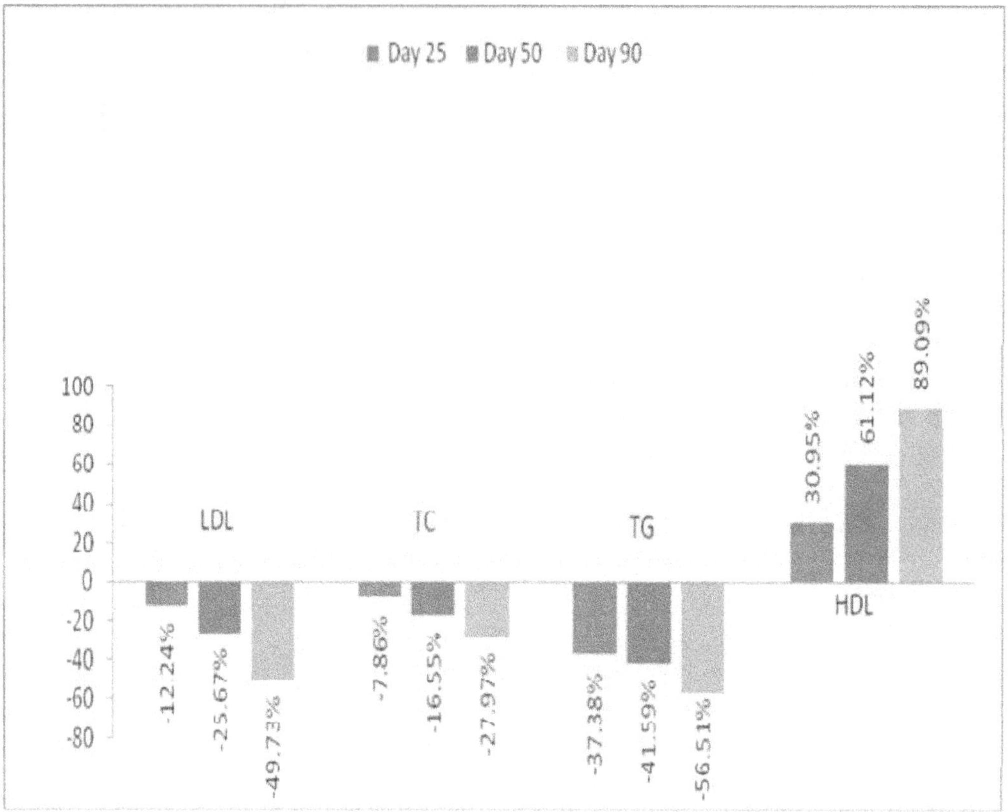

Figure 6.3.16 percentage Changes in Lipid Levels in Simvastatin, ezetimibe and Omega 3 Fatty Acids Group

PARAMETER-WISE OBSERVATIONS OF ALL GROUPS

Table 6.4 Serum Low Density Lipoproteins (LDL) cholesterol Levels

GROUP(G)	BASE VALUE	25TH DAY	50TH DAY	90TH DAY	%change
	mg/dl	mg/dl	mg/dl	mg/dl	0---90 days
G-I	133.0 ± 0.20	132.7 ± 0.17 (-0.22%)	132.5 ± 0.13 (-0.37%)	132.8 ± 0.20 (-0.15%)	0.15
G-2	181.5 ± 1.99	173 ± 1.97 (-4.68%)*	159.2 ± 1.95 (-12.28)*	150.1 ± 2.18 (-17.30)*	-17.3
G-3	200.1 ± 2.69	190.5 ± 2.50 (-4.79%)	176.8 ± 2.86 (-11.64%)*	161.7 ± 3.31 (-19.19)*	-19.19
G-4	180.1 ± 2.80	172.2 ± 3.18 (-4.38)	165.6 ± 3.15 (-8.05%)*	162.8 ± 2.78 (-9.60%)*	-9.6
G-5	185.6 ± 1.52	165.3 ± 1.79 (-10.93%)*	144.4 ± 1.73 (-22.19%)*	107.7 ± 1.79 (-41.97%)*	-41.97
G-6	196.7 ± 2.56	182.1 ± 2.13 (-7.42%)*	166 ± 2.25 (-15.60%)*	147.9 ± 2.60 (-24.80%)*	-24.8
G-7	184 ± 2.20	176.6 ± 2.06 (-4.02%)*	170 ± 1.64 (-7.60%)*	164.7 ± 2.26 (-10.48%)*	-10.48
G-8	192.8 ± 3.48	169.2 ± 3.66 (-12.24%)*	143.3 ± 2.67 (-25.67%)*	96.91 ± 2.78 (-49.73%)*	-49.73

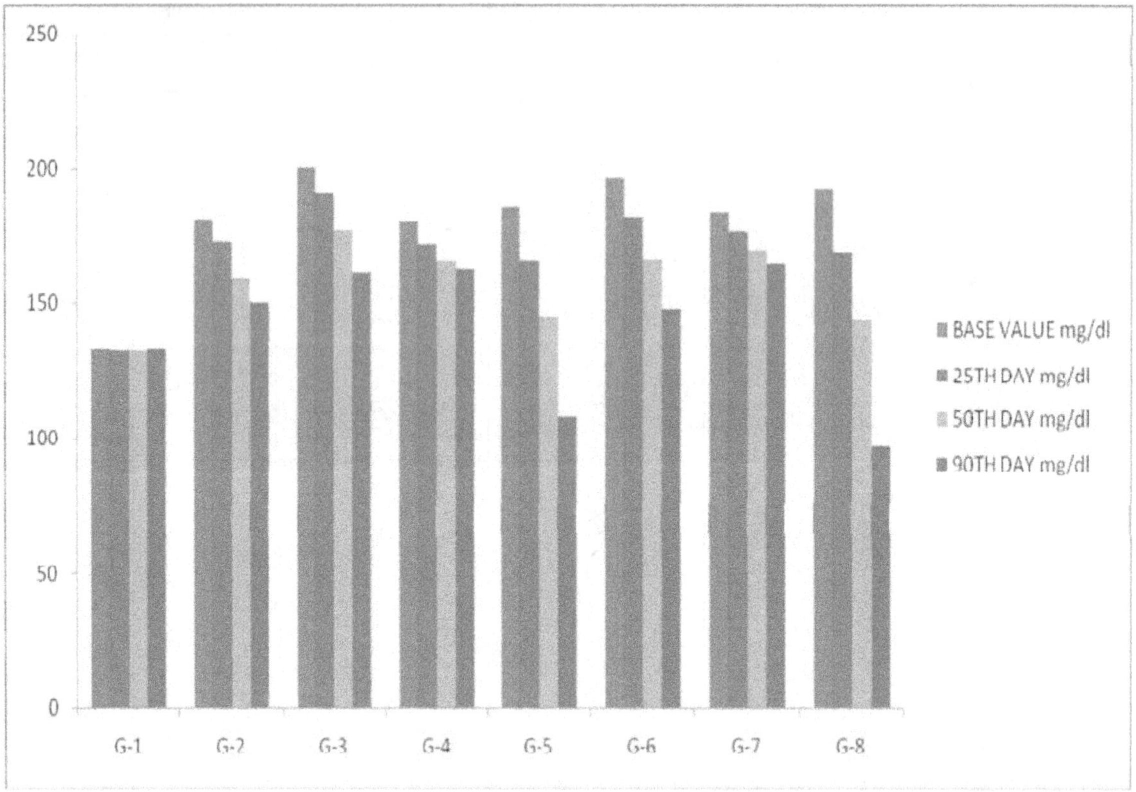

Figure 6.4.1 Serum LDL Levels

PARAMETER-WISE OBSERVATIONS OF ALL GROUPS

Table 6.5 Serum Total Cholesterol (TC) Levels

GROUP(G)	BASE VALUE	25TH DAY	50TH DAY	90TH DAY	%change
	mg/dl	mg/dl	mg/dl	mg/dl	0---90 days
G-I	232.4 ± 0.11	232.9 ± 0.30 (-0.21%)	232.8 ± 0.25 (-0.17%)	232.4 ± 0.26 (0%)	0
G-2	235.9 ± 2.23	228.4 ± 2.06 (-3.17%)	218.5 ± 2.06 (-7.37%)*	202.3 ± 2.62 (-14.24%)*	-14.24
G-3	227.2 ± 1.65	222.2 ± 1.48 (-2.20%)	215.3 ± 1.63 (-5.23%)*	200.5 ± 1.66 (-11.75%)*	-11.75
G-4	232.1 ± 2.04	225.5 ± 2.09 (-2.84%)	217.6 ± 1.98 (-6.24%)*	214.1 ± 2.16 (-7.75%)*	-7.75
G-5	234.1 ± 1.95	221.3 ± 2.16 (-5.46%)*	205.7 ± 2.96 (-12.13%)*	191.5 ± 2.52 (-18.19%)*	-18.19
G-6	234.5 ± 2.11	227.5 ± 2.24 (-2.98%)	219 ± 2.19 (-6.60%)*	204.7 ± 1.86 (-12.70%)*	-12.7
G-7	236.1 ± 1.51	228.3 ± 1.70 (-3.30%)*	218.7 ± 1.73 (-7.36%)*	210.2 ± 1.63 (-10.96%)*	-10.96
G-8	230.2 ± 2.15	212.1 ± 2.13 (-7.86%)*	192.1 ± 2.71 (-16.55%)*	165.8 ± 2.48 (-27.97%)*	-27.97

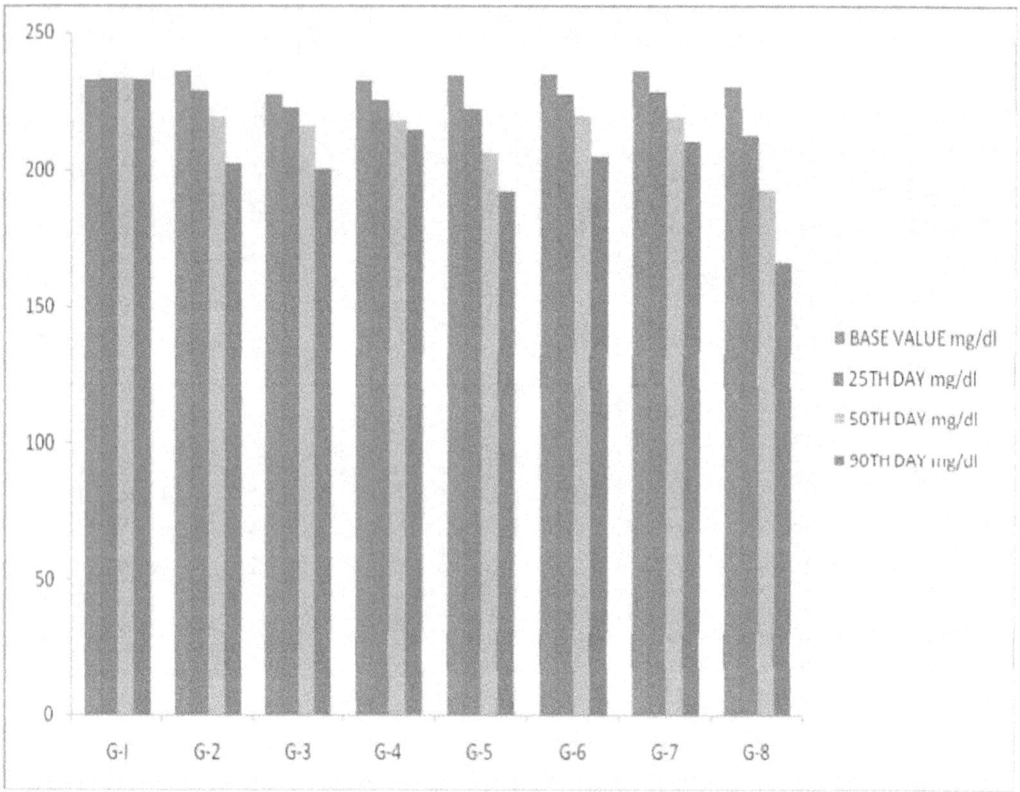

Figure 6.5.1 Serum Total Cholesterol (TC) Levels

PARAMETER-WISE OBSERVATIONS OF ALL GROUPS

Table 6.6 Serum Triglycerides (TG) Levels

GROUP(G)	BASE VALUE	25TH DAY	50TH DAY	90TH DAY	%change
	mg/dl	mg/dl	mg/dl	mg/dl	0---90 days
G-I	287.0 ± 0.17	287.0 ± 0.06 (0%)	286.7 ± 0.22 (-0.10%)	287.4 ± 0.18 (-0.13%)	0.13
G-2	230 ± 2.78	218.3 ± 3.20 (-5.08%)*	204.8 ± 2.77 (-10.95%)*	184.1 ± 2.36 (-19.95%)*	-19.95
G-3	243.5 ± 1.75	235.2 ± 1.71 (-3.40%)*	226.3 ± 2.08 (-7.06%)*	214.9 ± 1.82 (-11.74%)*	-11.74
G-4	231.7 ± 4.41	218.9 ± 4.07 (-5.52%)	204.8 ± 4.51 (-11.60%)*	197.5 ± 5.11 (-14.76%)*	-14.76
G-5	253.5 ± 3.39	226.9 ± 4.10 (-10.49%)*	200.6 ± 4.16 (-20.865)*	172.2 ± 1.98 (-32.07%)*	-32.07
G-6	222.2 ± 4.55	196.5 ± 3.99 (-11.56%)*	168.7 ± 4.36 (-24.07%)*	130.9 ± 4.08 (-41.08%)*	-41.08
G-7	264.7 ± 1.83	252.4 ± 1.75 (-4.64%)*	236.0 ± 1.37 (-10.84%)*	223.2 ± 1.82 (-15.67%)*	-15.67
G-8	280.3 ± 1.21	175.5 ± 1.25 (-37.38%)*	163.7 ± 1.61 (-41.59%)*	121.9 ± 1.30 (-56.51%)*	-56.51

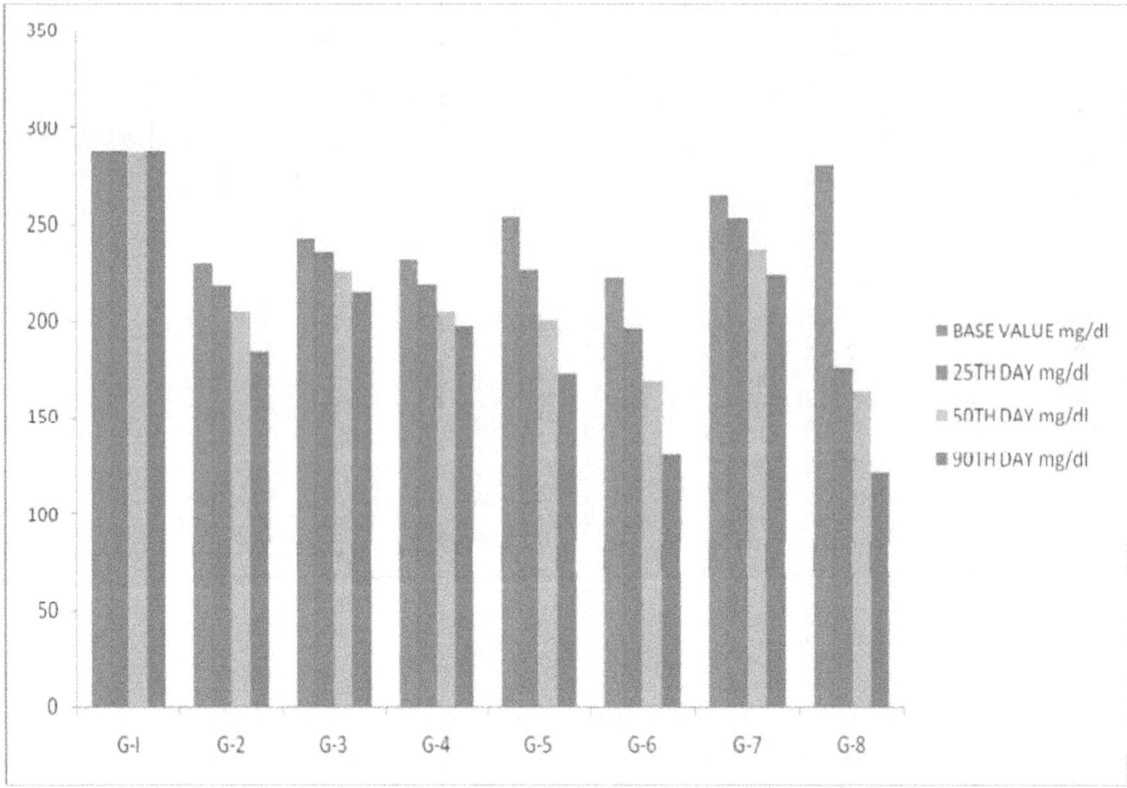

Figure 6.6.1 Serum Triglycerides (TG) Levels

PARAMETER-WISE OBSERVATIONS OF ALL GROUPS

Table 6.7 Serum High Density Lipoproteins (HDL) cholesterol Levels

GROUP(G)	BASE VALUE	25TH DAY	50TH DAY	90TH DAY	%change
	mg/dl	mg/dl	mg/dl	mg/dl	0-90 days
G-I	33.19 ± 0.16	32.76 ± 0.21 (-0.99%)	32.62 ± 0.10 (-1.31%)	33.00 ± 0.19 (-0.43%)	-0.43
G-2	33.95 ± 0.45	35.7 ± 0.34 (5.15%)*	37.50 ± 0.31 (10.45%)*	40.6 ± 0.44 (19.58%)*	19.58
G-3	33.95 ± 0.39	35.59 ± 0.35 (4.83%)*	36.77 ± 0.3 (8.30%)*	39.45 ± 0.39 (16.20%)*	16.20
G-4	34.1 ± 0.57	35.6 ± 0.44 (4.39%)	37.1 ± 0.43 (8.79%)*	40.7 ± 0.51 (19.35%)*	19.35
G-5	33.6 ± 0.52	35.75 ± 0.48 (6.39%)*	38.2 ± 0.45 (32.07%)*	42.5 ± 0.45 (26.48%)*	26.48
G-6	34.18 ± 0.45	35.91 ± 0.43 (5.06%)*	38.32 ± 0.34 (12.11%)*	42.36 ± 0.35 (23.93%)*	23.93
G-7	34.74 ± 0.32	35.83 ± 0.29 (3.13%)*	36.65 ± 0.30 (5.49%)*	38.04 ± 0.29 (9.49%)*	9.49
G-8	33.57 ± 0.54	43.96 ± 0.57 (30.95%)*	54.09 ± 0.7 (61.12%)*	63.48 ± 0.95 (89.09%)*	89.09

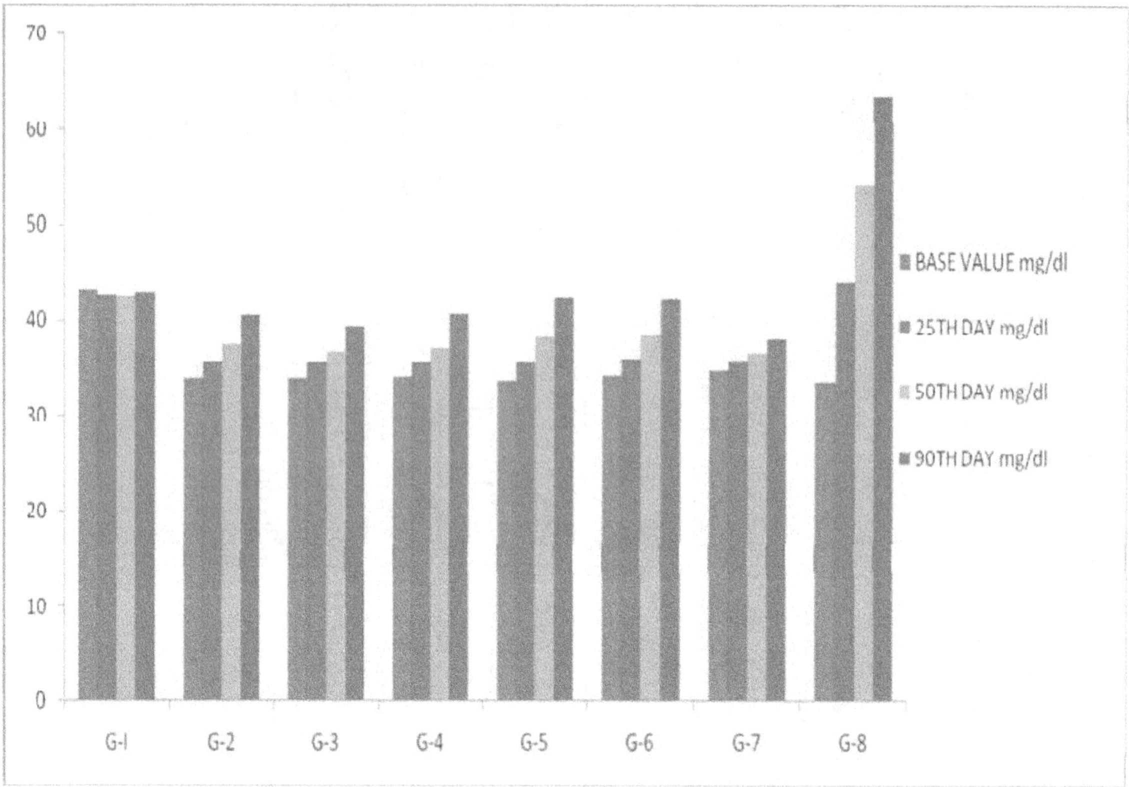

Figure 6.7.1 Serum HDL Levels

Table 6.8 Comparison of percentage change in lipid levels at the end of 25 days

Treatment	LDL decrease (% change)	TC decrease (% change)	TG decrease (% change)	HDL increase or changes (% change)
Group I	0.22	0.21	0	0.99
Group II	-4.68	-3.17	-5.08	5.15
Group III	-4.79	- 2.20	-3.40	4.83
Group IV	-4.38	-2.84	-5.52	4.39
Group V	-10.93	-5.46	-10.49	6.39
Group VI	-7.42	-2.98	-11.56	5.06
Group VII	-4.02	-3.30	-4.64	3.13
Group VIII	-12.24	-7.86	-37.38	30.95

(-) - Decrease , %- percentage

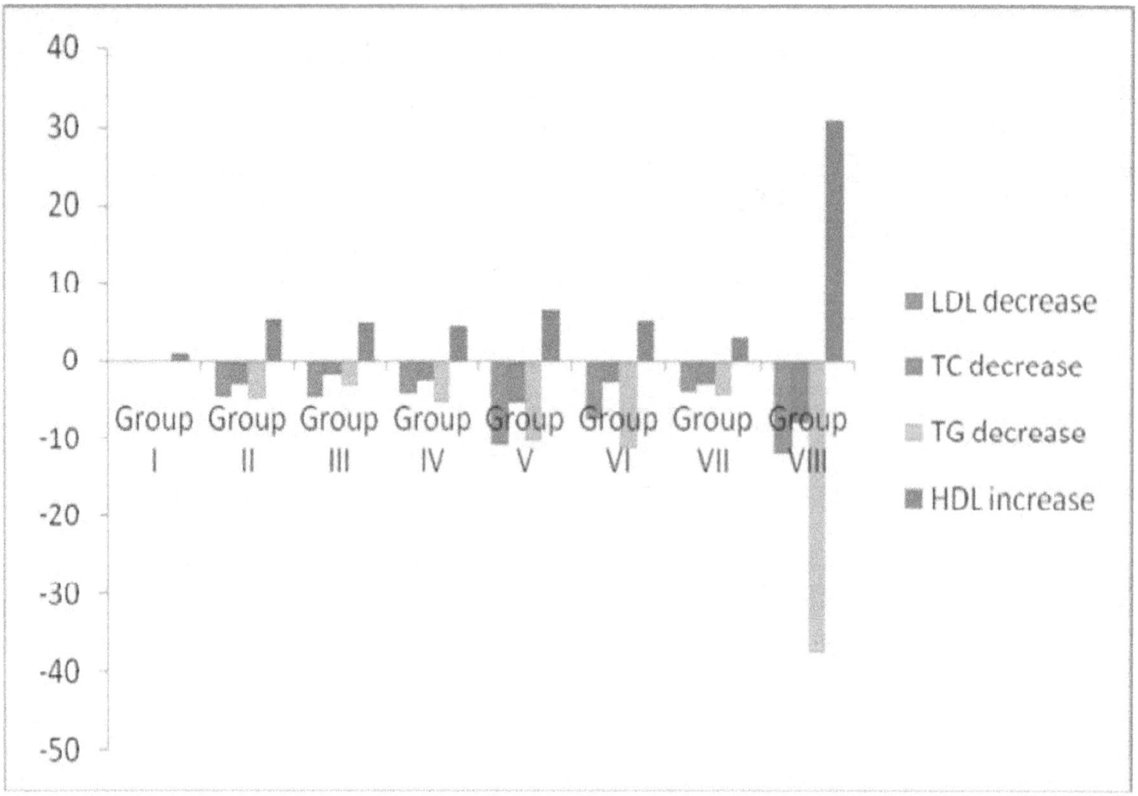

Figure 6.8.1 Comparison of percentage of change in lipid levels at the end of 25 days

Table 6.9 Comparison of percentage of change in lipid levels at the end of 50 days

Treatment	LDL decrease (% change)	TC decrease (% change)	TG decrease (% change)	HDL increase or change (% change)
Group I	0.37	0.17	0.10	1.31
Group II	-12.28	-7.37	-10.95	10.45
Group III	-11.64	-5.23	-7.06	8.30
Group IV	-8.05	-6.24	-11.60	8.79
Group V	-22.19	-12.13	-20.86	32.07
Group VI	-15.60	-6.60	-24.07	12.11
Group VII	-7.60	-7.36	-10.84	5.49
Group VIII	-25.67	-16.55	-41.59	61.12

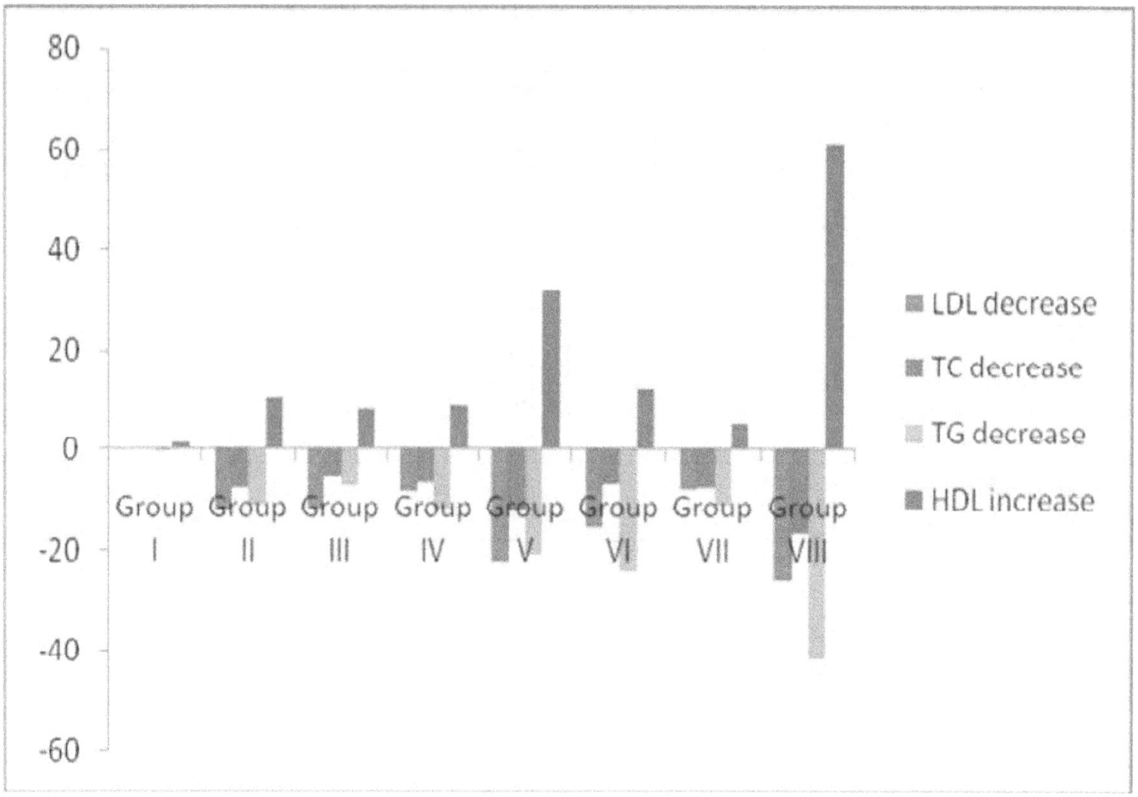

Figure 6.9.1 Comparison of percentage of change in lipid level at the end of 50 days

Table 6.10 Comparison of percentage change in lipid level at the end of 90 days

Treatment	LDL decrease (% change)	TC decrease (% change)	TG decrease (% change)	HDL increase or change (% change)
Group I	0.15	0	0.13	0.43
Group II	-17.30	-14.24	-19.95	19.58
Group III	-19.19	-11.75	-11.74	16.20
Group IV	-9.60	-7.75	-14.76	19.35
Group V	-41.97	-18.19	-32.07	26.48
Group VI	-24.80	-12.70	-41.08	23.93
Group VII	-10.48	-10.96	-15.67	9.49
Group VIII	-49.73	-27.97	-56.51	89.09

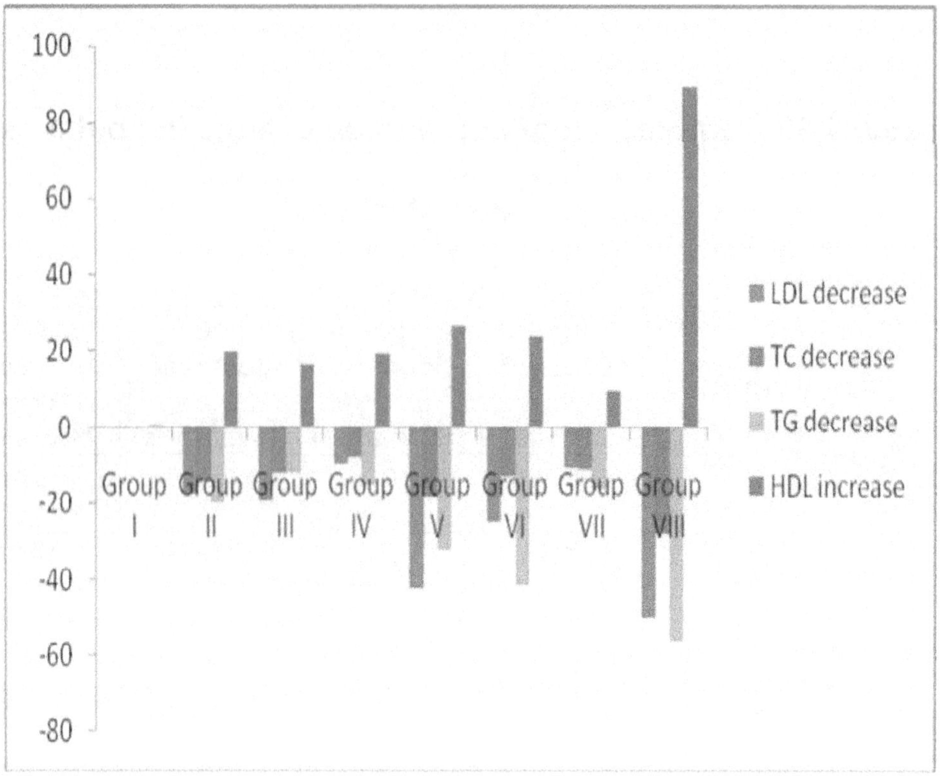

Figure 6.10.1 Comparison of percentage of change in lipid levels at the end of 90 days

7. CONCLUSIONS

Hyperlipidaemia is a major cause of atherosclerosis and atherosclerosis associated conditions, such as coronary heart disease (CHD), ischemic cerebrovascular disease, and peripheral vascular disease. These conditions still account for the majority of morbidity and mortality among middle-aged and older adults.

The reduction of elevated serum Total cholesterol, Triglycerides and low density lipoprotein cholesterol (LDL) and elevation of high density lipoprotein cholesterol (HDL) reduces the risk of coronary artery disease, resulting in a decrease in cardiovascular mortality.

Combination of drugs that act by different mechanisms can provide additive effects in reduction of LDL, TC and TG levels and elevation of HDL, useful to meet target levels. Combination therapy would be the desirable option to meet the target lipid levels in the management of hypercholesterolemia.

Statins are the most potent and frequently used drugs for the treatment of hypercholesterolemia. Statin therapy has been shown to reduce the rate of major vascular events in subjects with established vascular disease, and is considered to be the first line therapy for the management of dyslipidaemia in such individuals.

The co-administration of 10 mg of Ezetimibe and 20 mg of Simvastatin has the potential to produce clinically significant reductions in LDL, TC and TG level, comparable to Simvastatin or Ezetimibe alone, and with a favourable safety profile. Co-administration of Ezetimibe with a low-dose Simvastatin may be a safer, better-tolerated option than high-dose Simvastatin mono therapy for optimal cholesterol reduction. In another aspect, Simvastatin's cholesterol synthesis inhibition and ezetimibe's cholesterol absorption inhibition were the two different complementary mechanisms utilized to produce incremental decreases in blood LDL-cholesterol. Ezetimibe appears to lower blood triglyceride levels unlike resins, which increase blood triglycerides. Ezetimibe need not be taken with food and can be taken either in the morning or evening as a single 10 mg tablet. In addition, there are few drug–drug interactions. In high risk cases, Ezetimibe would be particularly useful for those subjects who cannot get to goal with current regimens. This treatment offers strategy to attain target levels of cholesterol that cannot be achieved with the highest doses of the Statin alone.

The combination therapy with Omega-3 fatty acids plus Simvastatin and/or Ezetimibe produced a greater reduction in triglyceride levels than simvastatin or ezetimibe mono therapy or Simvastatin and Ezetimibe combination therapy. A combination therapy of Omega-3-fatty acids with Simvastatin and/or Ezetimibe showed few adverse effects. Thus it could be considered as good therapeutic choice for subjects with mixed dyslipidemia, lowering triglycerides without mitigating LDL cholesterol reduction by statins or ezeimibe.

In a number of small studies, the combination of Statins or Ezetimibe and Omega-3 fatty acids have been consistently shown to be an effective, safe, and well-tolerated treatment for combined dyslipidaemia. Omega-3 fatty acids provide additional lipid improvements without requiring additional laboratory tests and do not increase risk for adverse muscle or liver effects.

Thus, a combination therapy of Omega-3 fatty acids plus Simvastatin and Ezetimibe would be considered an optimal treatment option for subjects with mixed dyslipidemia.

In conclusion, combined therapy of Ezetimibe 10 mg, Simvastatin 20 mg and 4 grams of Omega 3 fatty acids to subjects with hypercholesterolaemia was well tolerated and significantly reduced serum LDL, TG and TC levels and elevated HDL levels. Thus, combined therapy of Ezetimibe 10 mg, Simvastatin 20 mg and 4 grams of Omega 3 fatty acids is an alternative to titrating to higher doses of Simvastatin or Ezetimibe. Since mono therapy may be ineffective in reaching the target by Simvastatin, Ezetimibe or Omega-3 fatty acids, combination therapy would be the desirable option to meet the target lipid levels on the management of hypercholesterolemia.

The results from present study showed that the combined therapy of Ezetimibe, Simvastatin and Omega 3 fatty acids was well tolerated, with no evidence of increased incidence of adverse events or increases in clinical laboratory tests values indicative of liver or skeletal muscle toxicity.

Thus our study brings the following conclusions:

1) Among 171 subjects, 21 were between 20-30 years age group, 25 were between 31-40 years, 26 subjects between 41-50 years, 44 were between 51-60 years and 55 were more than 60 years age group.The prevalence of hyperlipidaemia was higher among 41-60 years age groups in study groups, showing more vulnerable age group for hyperlipidaemia. The study also highlights that younger age group is at increased risk of hyperlipidaemia.

2) Among 171 subjects, 101 were male subjects and 70 were female subjects. Males were more in number as compared to females in study groups.

3) The second drug, ezetimibe have a complementary lowering effect on LDL-C by atleast 19.19% with a single dose. Because of the nonlinear log dose-response curve of Statins, when the initial Statin dose is doubled, there is only an additional 6% lowering of LDL-C. Thus, a medication that would provide 19% lowering of LDL-C would save three doublings of the initial Statin dose to obtain the same effect on LDL-C.

4) When cholesterol synthesis was blocked by Simvastatin, the unwanted effects of Omega-3 fatty acids on total cholesterol and LDL cholesterol levels were kept under control in mixed

dyslipidemia subjects. In contrast to the potential serious side effects of combinations of Statins with fibrate, a combination therapy of Omega-3 fatty acids and Simvastatin showed few adverse events. Thus, it could be considered as good therapeutic choice for subjects with mixed dyslipidemia, lowering triglycerides by 41.08% without mitigating LDL cholesterol reduction by Statins.The triglycerides lowering efficacy was significantly increased with Simvastatin and Omega-3 fatty acids combination and with simvastatin, ezetimibe and Omega-3 fatty acids combination when compared with monotherapy of Simvastatin or Ezetimibe alone.

There were no serious adverse effects in the groups, in our study except for a few cases with complaints of allergic rashes, flatulence, constipation and diarrhea.

In conclusion, combined therapy of Ezetimibe 10 mg, Simvastatin 20 mg and Omega 3 fatty acids 4 g to subjects with mixed hypercholesterolaemia was well tolerated, significantly reduced serum LDL-C, TG and TC.

Goals of future studies are to establish the efficacy and tolerability of this combination therapy with large populations with primary Hypercholesterolaemia.

REFERENCES

1. Hyperlipidemia-update and review. Pharmacy continuing education from W-F professional associates. 2013; 35(1).

2. Allender S, Scarborough P, Peto V, Rayner M, Leal J and Luengo-Fernandez R. 2008. European cardiovascular disease statistics, ed. European Heart Network.

3. WHO, cardiovascular diseases, Fact sheet number 317, 2013, www.who.int/mediacentre/factsheets/fs317/en/index.html.

4. K.Park "Park's Text book of preventive and social medicine", Banarsidas Bhanot publishers, 21st edition, (2011) 338-339.

5. Tewari S, Kumar S, Kapoor A, Singh U, Agarwal A and Bharti B B, "Premature coronary artery disease in north India: an angiography study of 1971 patients", Indian Heart Journal, **57** (2005) 311-318.

6. Sodipo O A, Abdulrahman F I, Sandabe U K and Akinniyi J A, "Drug Therapy for hyperlipidaemia (dyslipidaemia)-A review. Journal of Applied Pharmaceutical Science, **1** (2011) 1-6.

7. Misra A, Luthra K and Vikram N K, "Dyslipidemia in Asian indians: Determinants and significance", Journal of the Associations of Physicians of India, **52** (2004) 137-142.

8. Libby: Braunwald's Heart Disease: A Textbook of cardiovascular medicine, 8th edition (2007) 1081.

9. Stone N J, "Secondary causes of hyperlipidemia", Medical Clinics of North America, **78** (1994) 117-41.

10. Jane-Charles F, Melchior C.N, Erick S.G.S, John J.P.K and Patrick D, "Atherosclerosis: Evolving vascular biology and clinical implications. New risk factors for atherosclerosis and patient risk assessment", Circulation, **109** (2004) III-15-III-19.

11. Krauss R M, Eckel R H, Appel L J, Daniels S R, Erdman J W and Kris-Etherton P, "AHA scientific statement. AHA dietary guidelines", Circulation, **102** (2000) 2284-99.

12. Lichtenstein A H, Appel L J, Brands M, Carnethon M, Daniels S and Franch H A, "Diet and lifestyle recommendations revision 2006: a scientific statement from the American Heart Association Nutrition Committee", Circulation, **114** (2006) 82-96.

13. Peter O.K, "A review of lipid-modifying drugs used alone and in combination", Endocrinology, **5** (2005) 476-491.

14. Theodore F, Michael K, William I, James M, Helmut S and Andrew L, "Treatment of high-risk patients with ezetimibe plus simvastatin coadministration versus simvastatin alone to attain National Cholesterol Education Program Adult Treatment Panel III

low-density lipoprotein cholesterol goals", American Journal of Cardiology, **93** (2004) 1481-1486.

15. Gloria L V, Patrick T S M, Nilo B C, Neil B C, Neil F and Shinichi M, "Effects of adding fenofibrate (200 mg/day) to simvastatin (10mg/day) in patients with combined hyperlipidemia and metabolic syndrome", The American Journal of Cardiology, **91** (2003) 956-960.

16. Erkelens D W, "Combination drug therapy with HMG CoA reductase inhibitors and bile acid sequestrants for hypercholesterolemia", Cardiology, **77** (1990) 33-38.

17. Harris W S, Miller M, Tighe A P, Davidson M H and Schaefer E J, "Omega-3 fatty acids and coronary heart disease risk: clinical and mechanistic perspectives", Atherosclerosis, **197** (2008) 12-24.

18. Rosenson R S, "The rationale for combination therapy", American Journal of Cardiology, **90** (2002) 2K-7K.

19. Stone N, "Combination therapy: its rationale and the role of ezetimibe", European Heart Journal Supplements, **4** (2002) J19-J22.

20. Jeffrey G S, "A review of the rationale for additional therapeutic interventions to attain lower LDL-C when statin therapy is not enough", Current Atherosclerosis Reports, **14** (2012) 33-40.

21. Davis H R and Veltri E P, "Zetia: inhibition of Niemann-Pick C1 like 1 (NPC1L1) to reduce intestinal cholesterol absorption and treat hyperlipidemia", Journal of Atherosclerosis and Thrombosis, **14** (2007) 99-108.

22. Third report of the National Cholesterol Education Program (NCEP) expert panel on detection, evaluation and treatment of high blood cholesterol in adults (Adult Treatment Panel III). National Cholesterol Education Program, National Heart, Lung and Blood Institute, National Institutes of Health, NH publication no.02-5215 (2002) 111-6.

23. Goodman and Gillman's, The pharmacological basis of therapeutics, The Mcgraw-Hill companies, "Drugs used in the treatment of hyperlipoproteinaemias", 9th edition (1996) 875– 897.

24. Satoskar R S, Bhandarkar S D and Nirmala N R, Pharmacology and Pharmacotherapeutics, popular prakashan, 19th edition (2005) 578-585.

25. Todd C R, Scott V H, Christopher P F M and Peter J H J, "Combination drug-diet therapies for dyslipidemia", Translational Research, **155** (2010) 221- 227.

26. Jeffrey B R, Jose G J, Valdis P, Hella V, Rachid M and Mary E H, "Consistency of effect of ezetimibe /simvastatin compared with intensified lipid-lowering treatment strategies in obese and non-obese diabetic subjects", Lipids in Health and Disease, **12** (2013) 103.

27. Hayati M Y, Abbie L C, Ren D, Jennifer A W, Frances L N and Clifford P S, " Limited impact of 2g/day omega-3 fatty acid ethyl esters (omacor®) on plasma lipids and inflammatory markers in patients awaiting carotid endarterectomy", Marine Drugs, **11** (2013) 3569-81.

28. Charles J G, Naseer K, Muhammad R, Jagjit P, Zia K and Ping W, "Titrating lovaza from 4 to 8 to 12 grams/day in patients with primary hypertriglyceridemia who had triglyceride levels >500mg/dl despite Conventional triglyceride lowering therapy", Lipids in Health and Disease, **11** (2012) 143.

29. Sang-Hyun K, Min-Kyung K, Hae-Young L, Hyun-Jae K, Yong-Jin K and Hyo-Soo K, "Prospective randomized comparison between omega-3 fatty acid supplements plus simvastatin versus simvastatin alone in Korean patients with mixed dyslipidemia: lipoprotein profiles

and heart rate variability", European Journal of Clinical Nutrition, **65** (2011) 110-116.

30. Michel F, Qian D, Arvind S, Amy O J-L and Philippe B, "Low incidence of paradoxical reductions in HDL-C levels in dyslipidemic patients treated with fenofibrate alone or in combination with Ezetimibe or ezetimibe/simvastatin", Lipids in Health and Disease, **10** (2011) 212.

31. Leiter L A, Betteridge D J, Farnier M, Guyton J R, Lin J and Shah A, "Lipid –altering efficacy and safety profile of combination therapy with ezetimibe/statin vs.statin monotherapy in patients with and without diabetes: an analysis of pooled data from 27 clinical trials", Diabetes, Obesity and Metabolism, **13** (2011) 615-628.

32. Peter P T, "Drug Treatment of Hyperlipidaemia: a guide to the rational use of lipid-lowering drugs", Drugs, **70** (2010) 1363-1379.

33. Gianluca B, Carlo B G, Antonio E P, Cristina L G and Carlo M R, "Ezetimibe+simvastatin versus doubling the dose of simvastatin in high cardiovascular risk diabetics: a multicenter, randomized trial (the LEAD study)", Cardiovascular Diabetology , **9** (2010) 20.

34. Kater A-L A, Batista M C and Ferreira S RG, "Synergistic effect of simvastatin and ezetimibe on lipid and pro-inflammatory profiles in

pre-diabetic subjects", Diabetology and Metabolic syndrome, **2** (2010).

35. Thomas S, Michael R, Diane T, Aditi S, William T and Patrice G, "Changes in cholesterol absorption and cholesterol synthesis caused by ezetimibe and/or simvastatin in men", Journal of Lipid Research, **50** (2009) 2117-23.

36. Sheridan M H and Gillian M K, "omega-3 ethylester Concentrate: a review of its use in secondary prevention post-myocardial infarction and the treatment of hypertriglyceridaemia", Drugs, **69** (2009) 1077-1105.

37. Winkler K, Schewe T, Putz G, Odunc N, Chafer G and Siegel E, "Fluvastatin/fenofibrate vs. simvastatin/ezetimibe in patients with metabolic syndrome: different effects on LDL-profiles", European Journal of Clinical Investigation, **39** (2009) 465.

38. Heinz D, "Statins, fibrates, nicotinic acid, cholesterol absorption inhibitors, anion-exchange resins, omega-3 fatty acids: Which drugs for which patients?" Fundamental and clinical pharmacology. **23** (2009) 687-692.

39 Derosa G, Angelo A D, Franzetti I G, Ragonesi P D, Gadaletas G and Scalise F, "Efficacy and safety of ezetimibe/simvastatin association on non-diabetic and diabetic patients with polygenic hypercholesterolemia or combined hyperlipidemia and previously

intolerant to standard statin treatment", Journal of Clinical Pharmacy and Therapeutics, **34** (2009) 267-276.

40. Masuda D, Nakagawa-Toyama Y, Nakatani K, Inagaki M, Tsubakio-Yamamoto K and Sandoval J C, "Ezetimibe improves postprandial hyperlipidaemia in patients with type IIb hyperlipidaemia", European Journal of Clinical Investigation, **39** (2009) 689-698.

41. Gideon R H, Geesje M D-T, Leonie C v v-v D Z, Jobien K O and Frank L J V, "Lipid –lowering therapy does not affect the postprandial drop in high density lipoprotein-cholesterol (HDL-c) plasma levels in obese men with metabolic syndrome: a randomized double blind crossover trial", Clinical Endocrinology, **69** (2008) 870-877.

42. Anouk V D G, Cynthia C-J, Maud N V, Mieke D T, Claude G and Genming S, "Efficacy and safety of coadministration of ezetimibe and simvastatin in adolescents with heterozygous familial hypercholesterolemia", Journal of the American college of cardiology, **52** (2008) 1421-1429.

43. Constance C, Westphal S, Chung N, Lund M, McCrary S and Johnson-Levonas A O, "Efficacy of ezetimibe/simvastatin 10/20 and 10/40 mg compared with atorvastatin 20mg in patients with type 2 diabetes mellitus", Diabetes Obesity and Metabolism, **9** (2007) 575-584.

44 .Teddy K, Paul S, Amy O J-L, John F P, Arthur J B and Kevin B A, “Ezetimibe , A review of its metabolism, pharmacokinetics and drug interactions”, Clinical pharmacokinetics, **44** (2005) 467-494.

45. Susan B, Philip S , Leslie L, Lorenzo M, Ramachandran S and Enrico V, “Consistency in efficacy and safety of ezetimibe coadministered with statins for treatment of hypercholesterolemia in women and men”, Journal of women’s Health, **13** (2004) 1101-1107.

46. Marc E, Aled R, Steve D and Alan R, “Medical Lipid-Regulating Therapy. Current Evidence, Ongoing Trials and Future developments”, Drugs, **64** (2004) 1181-1196.

47. Jennifer M and Henry K, “Cytochrome P450 drug interactions within the HMG-CoA reductase inhibitor class. Are they clinically relevant?”, Drug safety, **26** (2003) 13-21.

48. Nordoy A, Bonaa K H, Nilsen H, Berge R K, Hansen J B and Ingebretsen O C, “Effects of simvastatin and omega-3 fatty acids on plasma lipoproteins and lipid peroxidation in patients with combined hyperlipidaemia”, Journal of Internal Medicine, **243** (1998) 163-170.

49. David W and John F, “Pharmacokinetic-pharmacodynamic drug interactions with HMG-CoA reductase inhibitors”, Clinical Pharmacokinetics, **41** (2002) 343-370.

50. Teddy K , Ingo M ,Enrico P V, Paul S, Bo Y and Yali Z, "Pharmacodynamic interaction between the new selective cholesterol absorption inhibitor ezetimibe and simvastatin", British Journal of Clinical pharmacology, **54** (2002) 309-319.

51. Jennifer A K, Filip M, Min G, Dan H and Jiri J F, "Reaching target lipid levels in patients at high risk of cardiovascular event: The experience of a canadian tertiary care lipid clinic", Central European Journal of Public Health, **15** (2007) 106-109.

52. Gilbert R T, "Management of dyslipidaemia", Heart, **90** (2004) 949-955.

53. Vijay N and Christie M B, "Combination therapy with Statins and Omega-3 Fatty Acids", American Journal of Cardiology, **98** (2006) 34-38

54. Bays H E, Tighe A P, Sadovsky R and Davidson M H, "Prescription omega-3 fatty acids and their lipid effects: physiologic mechanisms of action and clinical implications", Expert Review of Cardiovascular Therapy, **6** (2008) 391-409.

55. Reitman S and Frankel S, "A colorimetric method for the determination of serum glutamic oxaloacetic and glutamic pyruvic transaminases", American Journal of Clinical Pathology, **28** (1957) 56-63

56. Sidney B R, "Enzyme assays in diseases of the heart and skeletal muscle", Journal of Clinical Pathology, **24** (1970) 60-70.

57. Nicholas V C, Robert W L and Joseph H R, "The detetermination of glycogen in liver and muscle by use of anthrone reagent", The Journal of Biological Chemistry, **220** (1956) 583-593.

58. Qizilbash N, Jones L, Warlow C and Mann J," Fibrinogen and lipid concentrations as risk factors for transient ischaemic attacks and minor ischaemic strokes", British Medical Journal, **303** (1991) 605-609.

59. Friedewald W T, Levy R I and Fredrickson D S, "Estimation of the concentration of low-density lipoprotein cholesterol in plasma without use of ultracentrifuge", Clinical Chemistry, **18** (1972) 499-502.

60. Nazir A S, "LDL-Cholesterol lowering effect of different generic products of simvastatin compared to simvastatin in Pakistani hypercholesterolemic subjects-A randomized study", University of Karachi (2006).

61. George Y, Khalid Z A-S and Robert A H, "Hypertriglyceridemia: its etiology, effects and treatment", Canadian Medical Association Journal, **176** (2007) 1113-20.

62. Lefer A M, Scalia R and Lefer D J, "Vascular effects of HMG CoA-reductase inhibitors (statins) unrelated to cholesterol lowering: new concepts for cardiovascular disease", Cardiovascular Research, **49** (2000) 281-287.

63. Rubinstein A and Izkhakov E, "Statins: an effective anti-atherosclerosis therapy", The Israel Medical Association Journal, **4** (2002) 456-457.

64. Catapano A L, Davidson M H, Ballantyne C M, Brady W E, Gazzara R A and Tomassini J E, "Lipid-altering efficacy of the ezetimibe/simvastatin single tablet versus rosuvastatin in hypercholesterolemic patients", Current Medical Research and Opinion, **22** (2006) 2041-2053.

65. Arja T E, Alice H L, Dariush M, and David M H, "Fish intake is associated with a reduced progression of coronary artery atherosclerosis in postmenopausal women with coronary artery disease", American Journal of Clinical Nutrition, **80** (2004) 626-32.

66. Alzira A.S C, Ursula W P L and Raul A V, "Statin and fibrate associated myopathy: study of eight patients", Arquivos De Neuro-Psiquiatria, **62** (2004) 257-261.

67. Ben J, Graham B and Michael D F, "Statins and peripheral neuropathy: causation or coincidence?", The British Journal of Diabetes and vascular Disease, **11** (2011) 38-39.

68. Bradford R H, Shear C L, Chremos A N, Dujovne C A, Franklin F A and Grillo R B, "Expanded clinical Evaluation of lovastatin (EXCEL) study results: two-year efficacy and safety follow-up", Americal Journal of Cardiology, **74** (1994) 667-73.

69. Bernhard R, Rudolf S, Steffen S, Frank D, Norbert V and Helmut G, " Omega, a randomized, placebo-controlled trial to test the effect of highly purified omega-3 fatty acids on top of modern guideline-adjusted therapy after myocardial infarction", Circulation, **122** (2010) 2152-2159.

70. Brown G, Albers J J, Fisher L D, Schaefer S M, Lin J T and Kaplan C, "Regression of coronary artery disease as a result of intensive lipid-lowering therapy in men with high levels of apolipoprotein B", The New England Journal of Medicine, **323** (1990) 1289-1298.

71. Bulcao C, Giuffrida F M A, Ribeiro-Filho F F and Ferreira S R G, "Are the beneficial cardiovascular effects of simvastatin and metformin also associated with a hormone-dependent mechanism improving insulin sensitivity?", Brazilian Journal of Medical and Biological Research, **40** (2007) 229-235.

72. Chronic disease management and prevention network. An Alberta Health capacity building initiative, Chinook health region Lethbridge, Alberta Vascular Protection Dyslipidaemia. Clinical Guide: (2006) 15-16.

73. Dawn E T, Brian G S, Jane Y E, Joseph D A, Barrett P H R and Murray W.H, "The molecular mechanisms underlying the reduction of LDL apoB-100 by ezetimibe plus simvastatin", Journal of Lipid Research, **48** (2007) 699-708..

74. Frank B H, Eunyoung C, Kathryn M R, Christine M A and JoAnn E M, "Fish and long-chain omega-3 fatty acid intake and risk of coronary heart disease and total mortality in diabetic women", Circulation, **107** (2003) 1852-1857.

75. Gonzalez-Ortis M, Martinez-Abundis E, Kam-Ramos A M, Hernandez-Salazar E and Ramos-Zavala M G, "Effect of ezetimibe on insulin sensitivity and lipid profile in obese and dyslipidaemic patients", Cardiovascular Drugs and Therapy, **20** (2006) 143-146.

76. Grundy S M, Vega G L, McGovern M E, Tulloch B R, Kendall D M, and Fitz-Patrick, "Efficacy, safety and tolerability of once-daily niacin for the treatment of dyslipidemia associated with type 2 diabetes: results of the assessment of diabetes control and evaluation of the efficacy of niaspan trial", Archieves of Internal Medicine, **162** (2002) 1568-1576.

77. Haffner S M, “Management of dyslipidemia in adults with diabetes”, Diabetes care, **21** (1998) 160-178.

78. Harry R D and Enrico P V, “Zetia: Inhibition of Niemann-pick C1 like 1(NPC1L1) to reduce intestinal cholesterol absorption and treat hyperlipidemia”, Journal of Atherosclerosis and Thrombosis, **14** (2007) 99-108.

79. Harsh Mohan, “Text book of pathology”, 5th edition, chapter 11: The blood vessels and Lymphatics, 280-281.

80. Hiroyasu I, Minatsu K, Junko I, Satoshi S, Katsutoshi O and Yoshikuni K, “Intake of fish and omega-3 fatty acids and risk of coronary heart disease among Japanese”, Circulation, **113** (2006) 195-202.

81 .Hermando L, Marcelo C S, Soori S, Marlene D, Trish C and Ross T T, “Effect of fish oil on arrhythmias and mortality: systematic review”, British Medical Journal, **337** (2008) a2931.

82. Irene G, Evangelos N L, Vasilios G S and Moses E, “Beneficial effects of omega-3 fatty acids: The current evidence”, Hellenic Journal of Cardiology, **47** (2006) 223-231.

83. Ipek M, Ozgur B, Pinar K K, TulinT, “Myopathy due to concominant use of statin and Gemfibrozil in a patient with chronic renal failure: case report”, Marmara medical journal, **17** (2004) 129-132.

84. Griffin J P, D'Arcy P F and Speirs C J, "A manual of adverse drug interactions", Butterworth & co.Ltd, London, 4th edition (1988) 03-61.

85. James S, Philip B, Rafel C, Deedwania P, Jean-Charles F and Steven H, "Effect of lowering LDL cholesterol substantially below currently recommended levels in patients with coronary heart disease and diabetes .The treating to new targets (TNT) study", Diabetes Care, **29** (2006) 1220-1226.

86. John C L, "Means and ends of statins and low-density lipoprotein cholesterol lowering", Journal of the American college of cardiology, **50** (2007) 419-420.

87. John J P K, Kevin C M, Andrey S, Marat E, Borge G N and Ben N M, "Omega-3 free fatty acids for the treatment of severe hypertriglyceridemia: The Epanova for lowering very high triglycerides (EVOLVE) trial", Journal of Clinical Lipidology, **8** (2014) 94-106.

88. Jan L B, "N-3 fatty acids and cardiovascular disease", The American Journal of Clinical Nutrition, **83** (2006) 1477S-1482S.

89. Jiangtao L, Bifeng W, Tianming X, Zhong L and Junzhu C, "Efficacy and tolerability of adding coenzyme A 400 U/d capsule to stable statin therapy for the treatment of patients with mixed dyslipidemia:

an 8-week, multicenter, double-blind randomized placebo-controlled study", Lipids in Health and Disease, **13** (2014).

90 .Guyton J R, Brown B G, Fazio S, Polis A, Tomassini J E and Tershakovec A M, "Lipid-altering efficacy and safety of ezetimibe/simvastatin coadministered with extended-release niacin in patients with type IIa or type IIb hyperlipidemia", Journal of the American college of cardiology, **51** (2008) 1564-72.

91. Jean D, "Beneficial cardiovascular pleiotropic effects of statins", Circulation, **109** (2004) III-39-43.

92. James K L, "Effects of statins on 3-hydroxy-3-methylglutaryl coenzyme A reductase inhibition beyond low-density lipoprotein cholesterol", American Journal of Cardiology, **96** (2005) 24F-33F.

93. Mehta J L, "Statins and altered glucose metabolism, a laboratory curiosity or a new disease?", Journal of the American College of Cardiology, **55** (2010) 1209-16

94. Kastelein J J, Akdim F, Stroes E S, Zwinderman A H, Bots M L and Stalen-hoef A F, "Simvastatin with or without ezetimibe in familial hypercholesterolemia", The New England Journal of Medicine, **358** (2008) 1431-43.

95 . Kevin C M, JamesMckenney, Barry C.L and Matthew S R, "Lipid effects of prescription omega-3 acid ethyl esters plus simvastatin in subjects with hypertriglyceridemia", The Journal of the Federation of American Societies for Experimental Biology, **22** (2008) 147-148.

96. Kausik K R and Christopher P C, "The potential relevance of the multiple lipid-independent (pleiotropic) effects of statins in the management of acute coronary syndromes", Journal of the American college of cardiology, **46** (2005).

97. Kang J X and Leaf A, "Prevention of fatal cardiac arrhythmias by polyunsaturated fatty acids", American Journal of clinical Nutrition, **71** (2000) 202-207.

98. Lipsy R J, "Overview of pharmacologic therapy for the treatment of dyslipidaemia", Journal of Managed Care Pharmacy, **9** (2003) 9-12.

99. Michael H D, Thomas M, Robert B, Lorenzo M, Leslie J L and Alexandre P.L, "Ezetimibe coadministered with simvastatin in patients with primary hypercholesterolemia", Journal of the American college of Cardiology, 40 (2002).

100. Madsen T, Skou H A, Hansen V E, Fog L, Christensen J H and Toft E , "C-reactive protein, dietary n-3 fatty acids, and the extent of coronary artery disease", American Journal of Cardiology, **88** (2001) 1139-42.

101 . Michael A B, Robert P G, Christopher P C, Thomas A M, Andrew M T and Jennifer A W, "Evaluating cardiovascular event reduction with ezetimibe as an adjunct to simvastatin in 18,144 patients after acute coronary syndromes: Final baseline characteristics of the IMPROVE-IT study population", American Heart Journal, **168** (2014) 205-212.

102. Marco M, Rafael P, Felipe S-J, Franz K and Anthony Hackney, "Creatine supplementation: effects on blood creatine kinase activity responses to resistance exercise and creatine kinase activity measurement", Brazilian Journal of Pharmaceutical Sciences, **45** (2009).

PAPERS PUBLISHED

1. **W.Clement Atlee** and M.Vasudevan, "Study of drug-drug interaction on the Management of hyperlipidemic disease: Simvastatin and Ezetimibe", International Journal of Research and Development in Pharmacy and Life Sciences, **2** (2013) 383-386.

2. **W.Clement Atlee** and M.Vasudevan, "A clinical study of drug-drug and drug-food interactions on the management of hyperlipidemic disease: Simvastatin, Ezetimibe and Omega- 3-fatty acids", International Journal of Research and Development in Pharmacy and Life Sciences, **3** (2014) 943-948.

3. **W.Clement Atlee** and M.Vasudevan,"Clinical study of pharmacodynamic drug Interaction on the management of hyperlipidemic disease", International Journal of Bioassays, **3** (2014) 1762-1764.

4. **W.Clement Atlee** and M.Vasudevan, "Comparing the effect of monotherapies of hyperlipidemia over placebo treatment", International Journal of Drug Development and Research, **6** (2014) 68-76.

www.ingramcontent.com/pod-product-compliance
Lightning Source LLC
LaVergne TN
LVHW082334200926
848481LV00012B/3344